Fielding's
LEWIS AND CLARK TRAIL

Current Fielding Titles

Red Guides—updated annually

FIELDING'S BERMUDA AND THE BAHAMAS 1986
FIELDING'S CARIBBEAN 1986
FIELDING'S DISCOVER EUROPE: OFF THE BEATEN PATH 1986
FIELDING'S ECONOMY CARIBBEAN 1986
FIELDING'S ECONOMY EUROPE 1986
FIELDING'S EUROPE 1986
FIELDING'S MEXICO 1986
FIELDING'S SELECTIVE SHOPPING GUIDE TO EUROPE 1986

Blue Guides—updated as necessary

FIELDING'S AFRICA: SOUTH OF THE SAHARA
FIELDING'S ALL-ASIA BUDGET GUIDE
FIELDING'S EGYPT AND THE ARCHAEOLOGICAL SITES
FIELDING'S EUROPE WITH CHILDREN
FIELDING'S FAMILY VACATION GUIDE
FIELDING'S FAR EAST
FIELDING'S HAVENS AND HIDEAWAYS USA
FIELDING'S LEWIS AND CLARK TRAIL
FIELDING'S MOTORING AND CAMPING EUROPE
FIELDING'S SPANISH TRAILS IN THE SOUTHWEST
FIELDING'S WORLDWIDE CRUISES 2nd revised edition

Fielding's
LEWIS AND CLARK TRAIL

by
Gerald W. Olmsted

Fielding Travel Books
c/o William Morrow & Company, Inc.
105 Madison Avenue, New York, N.Y. 10016

Copyright © 1986 by Gerald W. Olmsted

Maps by Carto Graphics, San Francisco
    Susan Camp, Kerry Chambers

All rights reserved. No part of this book may be reproduced or utilized in any form or by any means, electronic or mechanical, including photocopying, recording or by any information storage and retrieval system, without permission in writing from the Publisher. Inquiries should be addressed to Permissions Department, William Morrow and Company, Inc., 105 Madison Ave., New York, N.Y. 10016.

Library of Congress Catalog Card Number: 85-81756
ISBN: 0-688-05374-2

Printed in the United States of America

First Edition

1 2 3 4 5 6 7 8 9 10

To Eunice

ACKNOWLEDGMENTS

The author would like to thank Marjorie Fishman, Pat Heller, Curt Johnson, Steve Kappler, Robert Lange, Madeline Mixer, Leonard Nelson, Marder Shay, Bob Singer, and Wilbur Werner.

CONTENTS

Author's Note   1
The Ozarks   6
The Corn Belt   25
The Great Lakes   58
The Missouri Breaks   104
The Yellowstone River   137
The Continental Divide   155
The Blackfeet Country   187
The Bitterroot Mountains   206
The Inland Empire   223
The Cascades   245
The Pacific Coast Range   263
Practical Advice   286
Bibliography   291
Index   295

LIST OF MAPS

Lewis and Clark Expedition—1804, 1805, 1806   4
The Ozarks   8
The Corn Belt—Missouri and Kansas   26
The Corn Belt—Iowa and Nebraska   46
The Great Lakes—South Dakota   59
The Great Lakes—North Dakota   77
The Missouri Breaks   106
The Continental Divide   156
Blackfeet Country   188
The Bitterroots   208
The Inland Empire   224
The Cascades   246

Meriwether Lewis: Lewis is depicted as the visionary leader. The Nez Perce called him Yo-me-kol-lick: "White Bearskin Folded." Statue by James Earle Frasier: courtesy of Missouri State Museum, Jefferson City, MO. Photograph by Greg Leech.

William Clark: Clark, the expedition's principal chronicler and cartographer is shown with notebook in hand. He later became territorial governor of Louisiana. Statue by James Earle Frasier: courtesy of Missouri State Museum, Jefferson City, MO. Photograph by Greg Leech.

AUTHOR'S NOTE

This volume traces the route of Meriwether Lewis and William Clark from St. Louis to the Pacific Ocean. The explorers, of course, came home again, so it would be possible to organize the material in the reverse order, to go from the Pacific to the Mississippi, but to do so would be to miss a great deal of the drama that unfolded as the two captains struggled against the unknown. I'm afraid modern-day travelers following the route from west to east will, therefore, have to cope with a certain amount of disjointedness, and will miss a bit of the excitement of discovery.

The captains chose different routes for a portion of their homeward journey, so one might argue that the text in those sections should follow them west to east. The primary emphasis of the book, however, is to provide the modern adventurer with a travel guide. To this end, it is better to be geographically consistent. With one exception, all the text is organized on the assumption that you are going from east to west—following the sun, as it were. Regrettably, this means that you follow Clark along the Yellowstone in a backward (in time) direction. The exception is when you float through the "wild and scenic" section of the upper Missouri River. Here you *have* to go downstream, so you must follow Lewis and Clark from west to east.

I hope this book will be read aloud—a passenger reading to the driver about the things they are seeing along the way. This won't always be easy. The journals of Lewis and Clark were originally published in an edited form by Nicolas Biddle, who put their writings in the third person and cleaned up spelling, punctuation, and grammatical errors. In the process some of the charm was lost, and it is difficult, reading Biddle, to get a real sense of what the captains were thinking about, to learn what was inside their heads, to use the modern vernacular. Fortunately, in 1905 Dr. Reuben Gold Thwaites published the journals in their unedited form, which I have used whenever practical. The reader will have to cope with some innovative spelling and at times will chuckle at the fractured English the explorers seemed to favor. I have, however, altered the punctuation slightly and deleted some extraneous words to make it more readable.

In the interest of accuracy, I have identified the source of the quotes: "Biddle" means the Nicolas Biddle edition of 1811; "Lewis" or "Clark" means the text is taken from Thwaite's eight-volume account of the expedition. In 1893 Eliott Coues (pronounced Cows) reprinted Biddle's work and added extensive and enlightening footnotes of his own. From time to time his work is cited. Other quotes are self-explanatory.

If you look at a map of Lewis and Clark's trek and compare it with a modern guide showing what to see and do on a westward journey, you're bound to ask why you or I would want to follow that route. They wandered all over creation, yet missed all the fun places to see, like South Dakota's Black Hills and, in Wyoming, what is now Yellowstone Park. One place they did visit, Pierre, South Dakota, isn't on the road to anywhere, yet it is the site of a young America's first encounter with the Indian nations of the far west, and therefore has an intrinsic interest aside from its scenic beauty. The reader who follows the trek from one end to the other will discover a different, nontouristy world; one which played a large role in America's Manifest Destiny, and the winning of the west. Here again, though, things might seem a bit disjointed. You learn about American history *in situ;* an historical event is not discussed until you are at the very place where it happened. This may upset historians because we don't follow the fur trade, for example, in a nice, orderly, temporal fashion. Still, I hope you will find this approach refreshing and, because of your proximity to the actual locale, will come to a better understanding of the forces and events that shaped America's westward migration.

In recent times certain factions have protested the use of the word Indian to describe the Native American. They have a point, but it is not my privilege to change a place name to, for example, the "Standing Rock Native American Reservation," especially when it is administered by the B.I.A., the Bureau of Indian Affairs. The Native Americans we will learn about on this trip were, foremost, tribesmen, living in distinct nations that had their own legends. They lived by their own cultural mores, and so, whenever possible, I refer to them by their tribal names.

I haven't felt the need for a prologue. The reasons things happened as they did seem to come out in due time. Nor have I felt the need to expound on the significance of the expedition; like most who grew up in this country, I had that drummed into my head in high school.

I would, however, like to comment on the sponsor of the expedition. Thomas Jefferson not only had the vision to see clearly that the young nation would expand into the west and that an exploration of this type would be necessary, but he also contributed greatly to its success by the simple expedient of writing down exactly what he wanted done. Today's business school students would be well advised to read his instructions to the captains; they are the epitome of a well-thought-out business plan. A later president, John F. Kennedy, recognized his genius. At a formal White House dinner, to which he had invited America's Nobel Laureates to celebrate with him, he posed this toast: "There hasn't been this much intellect gathered around this table since Thomas Jefferson dined alone."

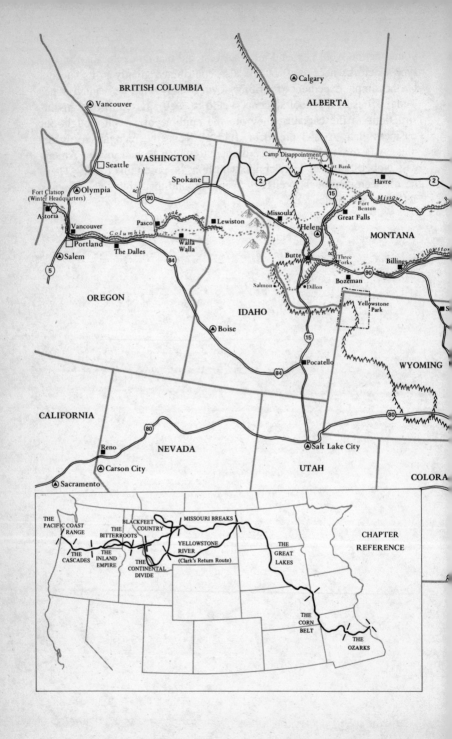

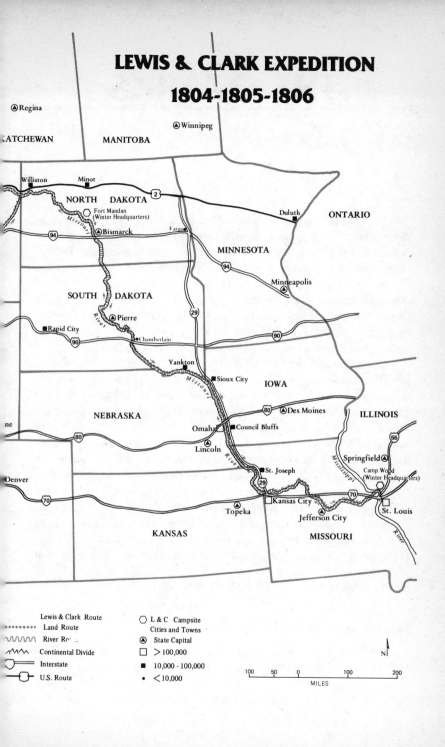

THE OZARKS

Our adventure begins in Illinois at the point where the Missouri meets the Mississippi. When Lewis and Clark arrived here in the waning months of 1803 this was the American frontier. Beyond the Mississippi lay France, or more accurately, country claimed by Napoleon's France. One hundred Native American nations had a stronger claim, but in 1803 the land ownership game was being played by European rules, and by those rules France held sovereignty over Louisiana. That, according to the thinking of the day, included "the Mississippi River and all its tributaries as far as their respective sources somewhere in the Stony Mountains."

Napoleon had halfheartedly ceded Louisiana to Spain—as a way to keep the British out—but Spain never committed the resources required to maintain control. Now Napoleon wanted out. He'd learned from an ill-fated adventure in San Domingo that maintaining an army in the New World was expensive and not very profitable. So he sold Louisiana to the United States for fifteen million dollars—four cents an acre. He needed the money to pay for his numerous other adventures, and this was a nice way to do so and still keep the area out of British hands.

While Lewis and Clark were camped in Illinois a treaty was signed in Paris that made Louisiana part of America. The land across the Mississippi River was foreign soil no more. Even so, the explorers, during their trek across the country and back, never quite accepted that fact. On numerous occasions their journals compare what they are seeing with what is found in the United States, and by the "United States" they clearly mean what's east of the Mississippi.

Missouri/Mississippi Confluence

Lewis and Clark arrived 130 years after the first Europeans sighted this meeting of the waters. In 1673, while drifting down the Mississippi, Marquette and Joliet heard a great roaring that they thought was produced by a rapid. They discovered instead the mouth of the Missouri. "I have seen nothing more dreadful," Marquette wrote. "An accumulation of large, entire trees, branches and floating islands we could not without great danger, risk passing through it."

Nearly 50 years later the French explorer Father Pierre Francois de Charlevoix commented: "I believe this is the finest confluence in the world. The two rivers are much the same breadth, each about half a league; but the Missouri is by far the most rapid, and seems to enter the Mississippi like a conqueror, through which it carries its white waters to the opposite shore without mixing them, afterwards, it gives its color to the Mississippi, which it never loses again but carries it quite down to the sea."

The Missouri is still muddy, even though the U.S. Army Corps of Engineers has built dam after silt-collecting dam upstream. "Too thin to plow, too thick to drink," goes the local saying. Forty tons of silt are deposited in the Louisiana delta every hour. It's also a fickle river. One hundred years ago the Sioux City (Iowa) *Register* commented tersely: "Of all the variable things in creation the most uncertain are the actions of a jury, the state of a woman's mind and the condition of the Missouri River." It would not take the captains long to learn how accurate these words were.

The Corps of Discovery, as they liked to call themselves, spent the winter in Illinois planning their great trek, training their men, organizing their stores, and dreaming of adventure. Man has always been fascinated by rivers. They are a living, dynamic thing; the thread that unifies the countryside, and that provides the means whereby the people who live along its shores can travel to other lands, to intercourse with other human beings. Where does the water come from? What is the source of this mighty stream; this artery that feeds our land? Questions like these no doubt taunted and excited the members of the Corps of Discovery, just as they did 75 years later, when men like Burton and Spekes and Stanley and Livingstone searched for the source of the Nile.

Today, if you look for the exact spot where the Corps of Discovery camped—that is, the exact latitude and longitude of the site—you'll find it's in Missouri, not Illinois, for both rivers have shifted course over the years. Furthermore, you'll find a swampland that looks nothing like the country where the explorers wintered. Today, look instead for the present confluence of the Mississippi and the Missouri and you'll then find yourself standing on the land that the two captains probably would recognize. Illinois' **Lewis and Clark State Park** lies directly across the Mississippi from where the muddy Missouri more than doubles the flow of the now combined rivers. The captains' winter campsite lay along

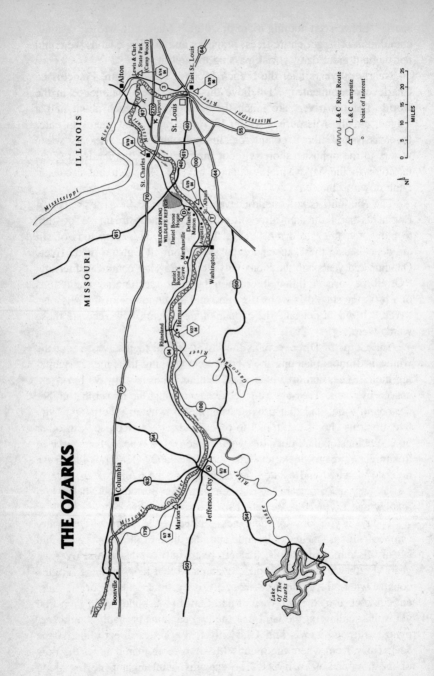

Wood (also called DuBois) River. Today Wood River is three or four miles farther north, near East Alton, Illinois. It flows through an oil refinery. By contrast, at Lewis and Clark State Park there is a small canal that resembles the former Wood River, so the illusion is one that exactly matches what the country looked like in 1803.

To reach the park, take State Highway 3 north from Interstate 270. About four miles south of the little town of Wood River, a dirt road leads past a quiet bayou where fishermen are trying their luck, passes through a forest of cottonwood, box elder, and sycamore, and continues on down to the Mississippi itself. A monument stands on a hummock at river's edge. Eleven concrete pillars, representing each of the eleven states the Corps of Discovery passed through on their way to the Pacific, have inscriptions describing some of the events along that momentous journey. In the spring, if the Mississippi is in flood, you must take off your shoes and wade out to the monument, which, upon reflection, proves to be a marvelous way of starting your own trek to the Pacific Ocean. You begin not at, or on, but *in* the Mississippi.

Technically, it's not quite true that the Lewis and Clark expedition started here on the left bank of the Mississippi. Meriwether Lewis spent much of the winter in St. Louis and didn't join William Clark and the rest of the party until they were 25 miles upstream. As the people of St. Charles, Missouri, will quickly tell you, Lewis & Clark, the pair, started in their fair town, and to emphasize the point, they will sell you a T-shirt that says St. Charles is "where the adventure began."

> *May 14th, 1804 [Biddle] The party consisted of: nine young men from Kentucky; 14 soldiers of the United States Army, two French watermen (Cruzatte, Labiche); an interpreter and hunter (Drouillard); and a black servant (York) belonging to captain Clark. All these except the last were enlisted to serve as privates during the expedition, and three sergeants appointed from amongst them by the captains (Ordway, Pryor, and Floyd). In addition to these were engaged a corporal and six soldiers, and nine watermen, to accompany the expedition as far as the Mandan nation.*

At 33, Clark was the oldest in the party. Lewis was four years his junior. John Shannon, 17, was the youngest. Many of the men would become heroes in their own right. Drouillard (the captains unaccountably spelled his name Drewyer) became one of the most respected mountain men in the west, a man whose life tragically ended at a place we will visit later on this odyssey. Two of the privates, Potts and John Colter, also became mountain men, the latter famous for his discovery of "Colter's Hell," a place we now call Yellowstone Park. One of the sergeants, Patrick Gass, lived in the limelight for a brief spell—his was the first journal published. Shannon, who later became a judge, was probably the only one intellectually equal to the captains. He later helped prepare the manuscripts for publication. Captain Lewis brought along his Newfoundland dog, Seaman. The composition of the party was to change at the Mandan villages, where two soldiers were sent home because of disciplinary problems and four people were added: a replacement soldier, the interpreter Toussaint Charbonneau, one of his wives, Sacagawea, and their infant son Baptiste, whom the captains nicknamed Pomp.

May 14th, 1804 [Biddle] The party was to embark on board of three boats; the first was a keelboat fifty-five feet long, drawing three feet of water, carrying one large squaresail and twenty-two oars. This was accompanied by two perioques (pirogues) or open boats, one of six and the other of seven oars. Two horses were at the same time to be led along the banks of the river for the purpose of bringing home game or hunting in case of scarcity.

Spring finally arrived, the ice-clogged river began to run free, and it was time to get moving.

May 14th, 1804 [Clark] Rained the fore part of the day I determined to go as far as St. Charles a french Village 7 Leags. up the Missourie, and wait at that place untill Capt. Lewis could finish the business in which he was obliged to attend to At St. Louis and join me by Land from that place 24 miles. I set out at 4 oClock P.M, in the presence of many of the neighbouring inhabitents, and proceeded on under a jentle brease up the Missourie.

The great adventure had begun. "We proceeded on" is a phrase written so often in the journals that it is now the title of the quarterly publication of the Lewis and Clark Trail Heritage Foundation.

Let us proceed on, too, but rather than heading directly for St. Charles, detour south and pick up Capt. Lewis's trail at St. Louis. Return south on Illinois Highway 3, but instead of crossing the river on Interstate 270, continue on along the Illinois side to U.S. Highway 40 and cross the river there. It's a dreary drive through miles and miles of railroad marshaling yards, but there is one unexpected pleasure. Along the way you get fine views of the "Arch," Eero Saarinen's masterpiece of sculpture that dominates the St. Louis skyline.

St. Louis

Largest of all the cities we will visit on this odyssey, St. Louis is also the oldest. The village was founded in 1763 by Pierre Laclede Liquest, a French fur trader who was looking for the first bit of dry ground he could find south of the swampland at the Missouri/Mississippi confluence. The town certainly didn't have a spectacular beginning. By the time Lewis and Clark arrived forty years had elapsed, yet only a thousand people had settled in the area. Now St. Louis boasts the 14th largest metropolitan population in the country with 2,377,000 souls living in the area.

The city is known principally for its beer and its baseball team, though not too long ago it was also the largest shoe manufacturing center in the U.S., giving rise to the saying "First in shoes, first in booze, and last in the National League." Later on men like Stan Musial would put the lie to that phrase. Shoe manufacturing left the country, but other industries stepped in, and now the city is home to some of America's largest corporations, including McDonnell Douglas, Monsanto, Anheuser-Busch, General Dynamics, and Ralston Purina.

Like many big cities in the U.S., St. Louis has had its problems. Its borders are too small—the city limits aren't very far from downtown; when urban decay set in the city started to lose the very people who could help, its more affluent population, who fled to separately chartered suburban towns. Today, in order to lure back those living in Clayton, Brentwood, and other outlying towns, the city is helping to build the St. Louis Centre, which will be the largest downtown shopping mall in the United States. The old railroad station, a masterpiece of 19th-century Romanesque architecture, is being refurbished along the lines of Boston's Faneuil Hall Marketplace. It's a fitting reuse of a building that, in its heyday, served 19 railroad companies and 260 trains a day.

During the winter of 1803–1804, Meriwether Lewis enjoyed the hospitality of the French fur trader Auguste Chouteau who, with his brother Pierre and their offspring, would dominate St. Louis' business and society for generations. For Lewis, this was a time of learning. The Chouteaus had traded on the Missouri with the Osages near present-day Kansas City, and were able to advise Lewis on that section of river. He studied maps drawn by the Spanish and the journals and seven charts drawn by the fur trapper John Evans, who had gone upriver as far as the Mandan villages in 1796–1797. Lewis suspected he would have a problem with the Teton Sioux, probably because he knew about the trading party conducted by one Baptiste Truteau ten years earlier. The Spaniard, Manuel Lisa, also shared what he knew about the upper river. During this period Lewis had time to ponder Jefferson's instructions:

> *June 20th, 1803 [Jefferson] The object of your mission is to explore the Missouri river, and such principal streams of it, as, by its course and communication with the waters of the Pacific ocean, whether the Columbia, Oregan, Colorado, or any other river, may offer the most direct and practicable water-communication across the continent for the purpose of commerce.*
>
> *You will take observations of latitude and longitude, at all remarkable points*
>
> *You will endeavour to make yourself acquainted with the names of the nations and their numbers;*
>
> *The extent and limits of their possessions;*
>
> *Their relations with other tribes;*
>
> *Their language, traditions, monuments;*
>
> *Their ordinary occupations*
>
> *Their food, clothing, and domestic accommodations*
>
> *Their diseases, and the remedies they use*
>
> *Moral and physical circumstances which distinguish them from the tribes we know;*
>
> *Peculiarities in their laws, customs, and dispositions*
>
> *Other objects worthy of notice will be—*
>
> *The soil and face of the country*
>
> *The animals of the country*
>
> *The remains and accounts of any which may be extinct*
>
> *The mineral productions*
>
> *Volcanic Appearances;*
>
> *Climate*

On March 10, 1804 both captains had the pleasure of watching the official transfer of Louisiana, first from Spain to France, and then, on the next day, from France to the United States. No trace exists of the Chouteau residence or the Spanish Government House, where the transfer took place. The one last surviving building of that era, Manuel Lisa's fur storehouse, was torn down to build the Gateway Arch. The bricks are stored in a St. Louis warehouse.

The prime attraction for Lewis and Clark buffs today, therefore, is a tour of the **Gateway Arch,** more formally known as the **Jefferson National Expansion Memorial,** operated by the National Park Service. In many ways this is a better place to start your Lewis and Clark trek than at the Missouri/Mississippi confluence because the excellent museum underneath the arch provides you with a preview of your trip.

Not surprisingly, many of the exhibits tell the story of our heroes' epic journey west. The museum centers on a lifesize statue of Jefferson, dressed in buckskins. All along the outer wall of the semicircular arena are huge photographs of some of the sights that the explorers saw, sights you will also encounter on your way west. Excerpts from the journals describe why each picture was chosen for the exhibit. Between Jefferson's statue and the outer wall are exhibits illustrating much of what happened between the time of their voyage and today, the varied things you will discover; you're presented with an outline—call it a syllabus—of the trip you are about to embark upon. Displays describe life among the tribes and illustrate the contributions of the mountain men, trappers, missionaries, cowboys, sod-busters, and railroad men, all of whom laid their imprimatur on this land. "Manifest Destiny" was the cry, and this museum is a tribute to that rather macho slogan. Still, the exhibits are done in such a sensitive way that those who abhor the way the government treated the native tribes can feel good about the way our western history is presented.

The arch itself is an exquisite piece of sculpture, taking on a different shape from each vantage point. Sunlight plays a moving picture on the polished stainless-steel skin, now adding, now deleting shimmering reflections of nearby buildings. It's a bit hard to take it all in without craning your neck. Little claustrophobia-inducing cars take you topside, where a short stairway leads to an observation room. The panorama from the base of the arch is breathtaking. Directly below is the river, where, on a fall day, a triumphant Corps of Discovery turned their craft toward shore for the last time.

September 23rd, 1806 [Clark] descended to the Mississippi and down that river to St. Louis at which place we arrived about 12 oClock. we Suffered the party to fire off thier pieces as a Salute to the Town. we were met by all the village and received a harty welcom from it's inhabitants &c.

The best way to view the arch, as well as another famous sight, the **Eads Bridge,** built in 1874, is to get out on the river. Two excursion boats, the *Huck Finn* and the *Tom Sawyer*, make a number of voyages each summer day. These are diesel-powered replicas of the stern-wheelers that used to ply these waters. The first steamboat to call at St. Louis was the *Zebulon M. Pike,* which arrived in 1817 after a six-week trip upriver from Louisville. More famous boats followed—the *Robert E. Lee* and the *Natchez,* which had their celebrated race upstream from New Orleans in 1870. If you have time, the best way to capture the flavor of that bygone era is to book passage aboard the *Delta Queen,* a genuine steamboat, or the newer, air-conditioned *Mississippi Queen,* both of which make multiday excursions throughout the summer.

Directly west of the arch is the **Old Courthouse,** where slave auctions were once held on the front steps. On a tour of the building you will learn that it was here Dred Scott started his fight for freedom, a sad part of our nation's history. The decision, which essentially upheld the right of one man to own another, so outraged much of the north that it hastened the outbreak of the Civil War. It's good to spend some time here because later on you'll be seeing battlefields from that great conflict as you follow Lewis and Clark's trail. In Missouri, the Civil War became "The War Within a State," rather than "The War Between the States." It was particularly savage here, literally pitting brother against brother.

Two other downtown attractions are the **Busch Stadium** and the **Anheuser-Busch facility,** the latter of which has daily tours followed by a sampling of the brewer's art. An old section along the river north of the arch has been redeveloped into a nightlife area called **Laclede's Landing,** where you'll get a taste of the Saint Louis Blues. Those who enjoy music with a more classical flavor can savor the sounds of the second-oldest symphony orchestra in the nation, and currently one of the best.

About 5 miles west of the Gateway Arch, along U.S. 40, is the immense **Forest Park,** where a number of attractions can keep you busy for days. The **St. Louis Zoo** is one of the finest in the country, with a miniature railroad to take you around the park. Nearby is the justly famous **St. Louis Art Museum,** housed in what was the Fine Arts Palace of the 1904 World's Fair. This was the only building constructed of permanent materials of the 1500 structures built to celebrate the centennial of the Lewis and Clark expedition. Outside, a statue depicts Saint Louis—Louis IX, King of France—sitting astride a mighty steed. On the north side of the park is another elegant structure where you can learn more about our two captains. The Missouri Historical Society's offices are in the **Jefferson Memorial Building,** which exhibits among other things, Meriwether Lewis's telescope and watch and one of William Clark's original elk skin–covered journals.

Forest Park abuts the city of *Clayton,* home of Washington University, one of the country's finest private schools. Nearby is the immense Romanesque-style **Cathedral of St. Louis,** with its Byzantine mosaic interior. Rare marbles, alabaster, and rose windows all add to the spiritual beauty of the church. A few miles south is the **Missouri Botanical Garden,** founded in 1859. Here you'll find the largest traditional Japanese garden in North America, with a 4-acre lake, islands, footbridges, waterfalls, lanterns, plum arbor, and teahouse.

Along the river south of town is **Jefferson Barracks Historic Park,** where the great chief Black Hawk was imprisoned, and where Robert E. Lee and Ulysses S. Grant both served for a time. Four of the original buildings have been restored and are open for public tours. These buildings are quite likely the only ones still extant that Captain Clark actually saw and used. President U.S. Grant spent his early married life at nearby **Hardscrabble Farm,** now owned by Anheuser-Busch and open to the public by advance reservation.

In the southwest corner of the metropolitan area is the **National Museum of Transport,** which boasts the largest collection of locomotives and rolling stock in the country. Featured is one of Union Pacific's famous Big Boys, the largest and most powerful locomotives ever built.

William Clark is buried in **Bellefontaine Cemetery,** located about 5 miles north of downtown. An obelisk and bust of the great explorer pays tribute not only to his exploits on his journey to the Pacific but also to his role as governor of the Missouri Territory, brigadier general in the territorial militia, and superintendent of Indian affairs. In the latter position, he once again demonstrated his genius. Under his wise tutelage the United States pursued an enlightened Indian policy that, sadly, was unappreciated at the time and came to an end after his death at age 69 in 1838. Also buried in Bellefontaine Cemetery are a number of men whose names will become familiar as you follow the trail of Lewis and Clark. Captain Bonneville is here, along with Senator Thomas Hart Benton, General Kearney, territorial governor of California, General Sterling Price of Civil War fame, the fur trader Manuel Lisa, and the famous mountain man Bill Sublette.

There is no memorial to Meriwether Lewis in Bellefontaine Cemetery. You have to go to Tennessee and travel the Natchez Trace Parkway to find his grave, which is located about 60 miles southwest of Nashville. There, on the night of October 10, 1809, Lewis, age 38, was either murdered or died by his own hand. Historians disagree: Jefferson believed he committed suicide; Bernard DeVoto and Lewis's biographer Richard Dillon both suggest it was murder.

St. Charles

Driving west from St. Louis on Interstate 70, you first pass the **Lambert International Airport,** where a replica of Lindbergh's *Spirit of St. Louis* hangs above the lobby, and then make the first of what will become your many crossings of the "Wide Missouri." Across the bridge is the only town besides St. Louis that was in existence in 1804.

May 16th, 1804 [Clark] we arrived at St. Charles at 12 oClock a number Spectators french & Indians flocked to the bank to See the party. This Village is about one mile in length, Situated on the North Side of the Missourie at the foot of a hill from which it takes its name Peetiete Coete *or the Little hill This Village Contns. about 100 houses, the most of them small and indeferent and about 450 inhabitents Chiefly French, those people appear Pore, polite & harmonious.*

Clark's French spelling is no better than his English.

The party spent five days here, readjusting the loads on their pirogues and the keelboat, and generally having a good time—a little too good a time for some, as it turned out. Clark had not quite perfected the military discipline he felt the corps needed and decided to set an example. He sentenced three men to run the gauntlet for being AWOL, drunk, and "behaveing in an unbecomeing manner at the Ball last night," a phrase that probably meant being a little too friendly with the ladies. He commuted the sentence of two, but forced the third (John Collins) to endure 50 lashes, dealt out by others in the party.

St. Charles has grown to a town of 40,000, most of whom commute into St. Louis or work at the airport. The area near the Interstate is rather plain, with a number of nice, though ordinary motels, but the old downtown is charming. An 8-block section has been restored to its turn-of-the-century appearance, with brick streets and buildings and quaint patios. Missouri's first capital, located on Main Street, is now a museum. Restaurants, gift shops, antiques stores, and galleries cater to tourists and bargain hunters. Washington Irving, who visited St. Charles in 1832, commented: "With its odd diminutive bowling green, skittle-ground, garden-plots, and arbours to booze in, it reminded us more of the Old World than anything we had seen for many weeks." He might say the same thing if he visited the town today.

The Missouri, Kansas, Texas Railroad (M.K.T. or Katy) runs between Main Street and the river, though a passenger train hasn't come through in 20 years. The old depot, restored to its nostalgic best, is now the **St. Charles Visitor Center.** The depot is set in a lovely riverside park, where you can sit on a bench and watch the tow boats go by. Fishermen talk about jug fishing for giant catfish. In jug fishing, you tie a 5-foot line to an empty gallon-size plastic milk carton, bait the hook with a chunk of liver, and throw the contraption overboard. Having repeated this process a dozen or so times, you then just float downstream, keeping your tackle always in sight. When the catfish takes the bait he tries to dive, but can't because of the milk jug. When he finally tires and gives up you simply paddle over and pick up your catch.

Daniel Boone Country

A bronze plaque along the St. Charles riverfront marks the spot where the Corps of Discovery finally got going. Snags and submerged logs called sawyers and crumbling riverbanks would plague the party for months, but on the whole, life was easy.

May 21st–22nd, 1804 [Clark] Set out at half passed three oClock under three Cheers from the gentlem on the bank. Soon after we came too, the Indians arrived with 4 Deer as a Present, for which we gave them two qts of Whiskey.

More than a year later, on the fourth of July, the Corps of Discovery—hot, dry, and with feet made sore by tramping over a cactus-strewn desert—drank their last drop of whiskey, and no doubt grew to resent that trade. These, incidentally, were the only natives the party would encounter until they reached present-day Council Bluffs, Iowa, more than two months later.

The route we follow begins on Boonslick Road, the oldest road west of the Mississippi. Originally called Boone's Lick Trace, it was built to haul salt from a manufacturing plant (salt lick) established in central Missouri by Daniel Morgan Boone. The western terminus was at Arrow Rock, where the Santa Fe Trail took off across the prairie. Like most roads of the day, it had problems. One Hampton Ball of Jonesburg, Missouri, who used it, recalled in his later years: "We did not always stick to the road. There were no fences. When one track became too muddy or rough with ruts, we drove out on the prairie or made a new road through the woods."

Cross the Interstate and proceed westward on State Highway 94. It's four lanes for a while, but after crossing U.S. Highway 40/61 it reverts to two lanes. A sign alongside the pavement lets you know you're on the right road; you'll see a silhouette of our heroes, one resting on his rifle, the other pointing the way, and below are the words: LEWIS AND CLARK TRAIL. You'll see a thousand of these signs on your westward trek, some of which are in the most isolated spots imaginable. They were erected by the various states at the suggestion of the Lewis and Clark Trail Commission, which was established by an act of Congress in 1964.

After a few miles you'll suddenly find yourself in the country, a refreshing change from the industry and bustle of St. Louis. The highway dips and turns and ducks into hollows, and then out onto riverbottom land where wheat grows in abundance. This is hilly country, the northern edge of the Ozarks. You pass through **Weldon Spring Wildlife Area,** whose trails lead out into forests of silver maple, dogwood, and box elder. Fox, squirrel, raccoon, white-tailed deer, beaver, skunk, and groundhog share the woods with 30 species of birds, including the great horned owl, great blue heron, marsh hawk, and kingfisher. A little farther along, turn right on county road D and proceed 5½ miles over gentle hills and through bayoulike riverbottoms to the house where Daniel Boone died. Old Dan and his wife Rebecca moved here when he was in his seventies because, as he put it: "Kentucky became too crowded—I could see smoke from my neighbors chimney." The Spanish government had given him land along the beautiful Femme Osage Creek, so named because an early French settler found the body of an Osage Indian woman in the stream. Dan was given the title of *syndic,* meaning judge. He decided disputes under a large elm tree called **The Judgment Tree,** the stump of which you can see today. The house, built by his sons, is now a privately run museum.

Did Lewis and Clark stop by to pay their respects to old Dan? The journals don't mention any meeting, so they probably did not, but the lady conducting the tours won't buy that argument. "It was no big deal," she asserts without a trace of doubt in her voice, "they just had a cup of coffee under the Judgment Tree."

Dan's grave is in nearby *Marthasville,* where recent evidence indicates that Missouri got the last laugh on Kentucky. Kentucky, it seems, felt Dan's remains belonged there and connived to have them moved, but, so the story goes, they got the wrong set of bones. The grave now in the bluegrass state is that of a former slave.

As you continue west, you pass through two hamlets, *Defiance* and *Matson.* Matson has an old, white clapboard store with two ancient gas pumps out front. The locals sit on a bench and wave at the people driving by. A little beyond Matson, the highway rejoins the river for a short while. On the opposite bank you see white limestone cliffs. Captain Lewis almost ended his adventure before it began; while he was exploring the country above those cliffs, he suddenly fell. He saved himself by implanting his espotoon (a spearlike weapon that he often carried) in the rock.

Nearby, Clark explored a local landmark known as **Tavern Cave.** The cave is still there, but it's farther from the river and substantially smaller due to filling of debris from the building of the Rock Island Railroad. To visit the cave, cross the river at Washington, take highway 100 east to county road T, go north to the hamlet of *St. Albans,* and follow an unnumbered road north to the river. You then cross the railroad at the mile 38 sign, slide down the hill about 60 feet, and poke around in the brush. What you find is really not worth the effort.

May 25th, 1804 [Clark] Camped at the mouth of a Creek called River a Chouritte, above a Small french Village of 7 houses and as many families, settled at this place to be convt. to hunt & trade with the Indians. The people at this Village is pore, houses Small, they sent us milk & eggs to eat.

These were the last white settlers the party would see and the last milk and domestic eggs they would eat for 2½ years. From this point on, the Corps of Discovery learned to live off the land. The site of this settlement, unfortunately, has been obliterated by the changing river.

As you continue west on Route 94, look for the turn-off to *Augusta.* This dreamy little town, with a white church steeple poking up above the trees, sits on a knoll overlooking the river. Augusta has the distinction of being the first *appellation controlle* district in North America, which means vintners can name their wine after the locality. You can tour two **wineries;** one overlooks the river and the other prides itself on being over 100 years old.

Augusta is the gateway to the German part of the state, the area that is often called the Missouri Rhineland. The hamlet of **Dutzow** reportedly was named by a German immigrant who, having been informed that the steamboat was about to leave, replied, "Dot So!" The town is so small you might be tempted to comment: "Where'd it go?" Cross the river here and proceed to the first substantial town west of St. Charles.

Washington

Washington is one of those old towns that has a lot of charm, but is not so cute that it has become touristy. Narrow streets line the hillside, which rises sharply from the grassy riverfront park. It's the kind of town

Norman Rockwell would have liked to paint. Many of the old red-brick houses were built flush with the sidewalk and display freshly painted shutters and simple brickwork cornices. Like most Missouri River towns, the business center is cut off from the water by the train tracks. Those with an eye for beauty feel ambivalent about the presence of the eyesore; they know it was the railroad that brought life to cities like Washington. The Missouri Pacific Railroad (Mo Pac) came on the scene in the 1850s. Its arrival spurred industry; in Washington's case, smoking pipes. Washington currently produces over 12 million corncob pipes a year. **Buescher's Cob Pipes factory** is open to tourists most workday mornings.

The 10,000 or so people who live here can enjoy the lovely park along the riverbank. On the bluffs above are a number of 1840s-style mansions. One of them, Elijah McLean's, is now a restaurant, and another is a bed-and-breakfast place.

You can proceed on to Hermann, our next stop, on either side of the river—both roads are lovely. Just west of Washington, on the south side of the river, is the farm where John Colter of Lewis and Clark and Yellowstone Park fame retired to a life he presumed would be safer than being chased by the Blackfeet. Like his friend George Drouillard, Colter played an epic role in the history of the fur trade. He married an Indian woman named Sally and settled here in 1810, only to die of pneumonia three years later. His grave was unwittingly destroyed during the building of the railroad.

Hermann

It's not difficult to discover the heritage of this Missouri town. Just look at the street names: Schiller, Goethe, Mozart. The festivals held in this little town also provide a hint: Maifest, Oktoberfest. At one restaurant you can enjoy an afternoon sojourn at the *Winestube mit fireplace und chairs mit gemutlich*. A stop at the visitor center in the German School building reveals that Hermann was founded in 1836 by the German Settlement Society of Philadelphia. Less than 3000 people live here now, and the town hasn't lost its old-world charm. White frame houses are tucked beneath 100-year-old elms. On a hot summer evening it seems as if half the population is sitting on their front porches watching the fireflies dance in the sky. The dome-topped schoolhouse, sitting atop a bluff overlooking the river, seems to be anxiously waiting for someone to pull the bell ropes to summon yet another class to its ancient doors.

Just as San Francisco has its Napa Valley, St. Louis has its Rhineland. Tours of the wine country include a stop at the **Hermannhoff Winery** for a sampling of the oenologist's art and the purchase of some wurst and cheese for a picnic on the lawn. The day might be capped off by dining at the Vintage 1847 restaurant at the **Stone Hill Winery,** a century-old building perched atop a knoll overlooking countless hills and dales.

For a while Hermann was a shipbuilding center, but there's no trace of it now. Building boats for the Missouri was more difficult than other rivers, according to author George Fitch:

The Missouri River steamboat should be shallow, lithe, deep-chested, and exceedingly strong in the stern wheel. It should be hinged in the middle and should be fitted with a suction dredge so that when it cannot climb over a sandbar it can assimilate it. The Missouri River steamboat should be able to make use of a channel, but should not have to depend on it. A steamer that cannot on occasion, climb a steep clay bank, go across a corn field, and corner a river that is trying to get away, has little excuse for trying to navigate the Missouri.

It's no wonder, then, that in Hermann today only a lookout high above the river recalls the glory days when steamboats were king.

The drive from Hermann to Jefferson City is beautiful, regardless of whether you're on the north or south side of the river. On the south side the road crosses the Gasconade River, where the explorers camped on June 27. This is the site of an early railroad tragedy. In their haste to get a line to Jefferson City, the railway builders constructed a flimsy trestle across the river. Then, compounding the misdeed, they decided to add some pomp to the festivities honoring the first train's arrival, so they stopped in Hermann to pick up a detachment of soldiers. Eight coaches plunged into the Gasconade River killing 28 people. The first train didn't reach Jefferson City for another year.

The road along the north side of the river goes through the hamlet of **Rhineland,** where a few dozen white gingerbreaddy houses sit amid lush green lawns beneath ancient elms. Someone, a proud papa, perhaps, has altered the road sign on the edge of town. "Population 172" has been crossed out—173 painted in.

Jefferson City

The Corps of Discovery continued struggling upstream. Pouring rain and adverse winds caused them to halt several days, and twice snags caused the boats to broach and nearly founder.

> June 2nd, 1804 [Clark] George Drewyer [Drouillard] and John Shields who we had sent with the horses Land on the N. Side, joined us this evening much worsted, they being absent Seven Days depending on their gun, the greater part of the time rain, they were obliged to raft or Swim many Creeks, those men gave a flattering account of the Countrey Several rats of Considerable Size was Caught in the woods to day.

They were approaching the present site of Jefferson City, a spot chosen for the state capital because of its central location and its high bluff, which would make a beautiful spot for a capitol building. This city of 40,000 has a big-city feel because state employees tend to dress up for work. The **capitol,** patterned after the one in Washington, DC, has an historical exhibit on the main floor and a second floor graced with statues of our heroes. Lewis, the visionary leader, stands tall with eyes fixed on the horizon, while Clark, as the expedition's cartographer, is shown writing in his journal. Outside, on the edge of the bluff, a large statue depicts the signing of the Louisiana Purchase in Paris. Guided tours of the capitol are given daily.

Besides the capitol, the principal tourist attraction is **Jefferson Landing,** a group of restored 1830s buildings operated by the State Department of Natural Resources. A museum is housed in a building used for a "boat stores." Note the word boat, not ship. Were Jefferson Landing on the ocean, the building would house a "ship chandlery." Across the street is the **Union Hotel,** now one of the most picturesque Amtrak stations in the country.

In spite of its name, Jefferson Landing is not on the river, and there is no landing. It's surprising that in a city picked to be the capital because of its river location there is no longer any place where you can even get to its banks, much less tie up a boat. The Army Corps of Engineers, in its effort to improve navigation and convert former floodplain into arable land, has completely altered the character of the land surrounding Jefferson City. Great levies and dikes were constructed in some places, but here lateral dams were built out into the river to force the water into a more narrow, faster channel, thus helping to scour out the bottom. The residue from the scouring filled the area between the dams and there, over the years, trees began to grow, creating a large wooded area. It's now a no-man's land that quite effectively isolates the buildings of Jefferson Landing from the river they were originally built to serve.

On their return voyage the Corps of Discovery must have sensed that the river they had more or less pioneered two years earlier was beginning to become a freeway. In a 3-week period, they met no less than 11 separate parties of trappers heading west. Excitement was building for what they knew would be a triumphant return.

> *Sept 17, 1806 [Clark] at 11 A.M. we met a Captain McClellin [who] informed us that we had been long Since given out by the people of the U. S. Generaly and almost forgotten, the President of the U. States had yet hopes of us*

> *Sept 20, 1806 [Clark] the party being extreemly anxious to get down ply their ores very well, we saw some cows on the bank which was a joyfull Sight to the party. every person, both French and americans seem to express great pleasure at our return, and acknowledge themselves much astonished in seeing us return, they informed us that we were supposed to have been lost long since.*

Marion

As you drive west from Jefferson's capitol building, Main Street soon joins Rock Hill Road, which then becomes State Highway 179. The road climbs a bluff where stately brick houses occupy beautiful sites overlooking the river. The vegetation is lush, as it was when the explorers were here: They feasted on strawberries, mulberries, purple raspberries, and wild currants.

> *June 8th, 1804 [Clark] Capt. Lewis went out above the river & proceeded on one mile, finding the countrey rich, the wedes & vines So thick & high he came to the Boat.*

Today, wild, flowering vines climb the telephone poles, and the shoulders of the road seem to be paved with beautiful orange wild flowers. You may have to slow down and swerve to avoid hitting a turtle inching across the road. The topography is marked by hills and hollows, by glens that become bayous, and is surrounded by sycamore trees and moss-covered white walnuts. This is the humid midwest, much like what the westward-bound traveler has been seeing all the way from Kentucky. It's good to take special notice of the countryside while you're here, because you are approaching a transition zone where you leave the Ozarks and enter the Corn Belt. The explorers, on their return from the high plains, noted the change in weather.

September 9th, 1806 [Clark] the climate is every day preceptably wormer and air more Sultrey than I have experienced for a long time.

The last "midwest" town you visit is Marion, not much more than a wide spot in the road, but one blessed with all the charm you've been finding since leaving St. Charles. Beyond Marion the road climbs up to the prairie, and, voila, you're now in the "west." The contrast is striking. You've seen corn and wheat fields before, but only along the river bottom. Here, they suddenly stretch out to the horizon, punctuated only by oak savannahs, which break the skyline into lovely artistic shapes. You have arrived at the first of a number of transition zones you will pass through on your trek to the Pacific. Behind you lies a country where settlers had to clear the land of trees so crops could be grown. Ahead, it was just the opposite. As settlements progressed westward, newcomers began to homestead land where there were no trees at all. One of a settlers' first tasks was to plant some. Aside from the cottonwoods along the river bottoms, virtually all the trees between here and the Rockies were planted to provide shade, firewood, or to give relief from the wind. It's nice to remember that the trees you will see on the next phase of your trip serve as a symbol of the heroic struggles the pioneers faced as they tamed the vast country ahead. You are about to enter the "Corn Belt."

THE CORN BELT

Early June found the Corps of Discovery entering prairie country. Captain Clark was quick to note the difference in the character of the land.

June 10th, 1804 [Clark] I walked out three miles, found the prarie composed of good Land and plenty of water roleing & interspursed with points of timber land. Those Pruries are not like those, E. Of the Mississippi, void of every thing except grass, they abound with Hasel Grapes & wild plumb of a Superior size & quality.

Little Dixie

The party was entering a country often referred to as Little Dixie. Pioneer families moving west from Kentucky, Tennessee, and Virginia knew a good spot when they saw one. They settled here with their slaves in the 1840s. It's a sleepy place now, dotted with a half-dozen towns that have a great deal of charm. Driving along a back road, you may see two farmers, clad in bib overalls, chatting next to the mail box. They wave when they see your out-of-state license plate. Downtown streets in the small towns are quiet, even at midday; you can hear yourself talk when using a curbside pay phone. The social center is the sandwich counter at the drugstore. Middle-aged ladies wearing cotton dresses and double-knit slacks serve coffee in cups with saucers and a tiny glass bottle of real milk is on the side. Those damnable triangular-shaped cartons of nondairy creamer haven't found their way to this part of the world.

When the explorers were in this part of the country they seemed to be having their share of problems. On June 4 Clark wrote: "The Serjt at the helm run undr a bending Tree & broke the Mast." Sergeant Ordway, who also kept a journal, candidly confessed: "Our mast broke by my Stearing the Boat near the Shore the Rope or Stay to her mast got fast in a limb of a Secamore tree & it broke verry Easy." The next day

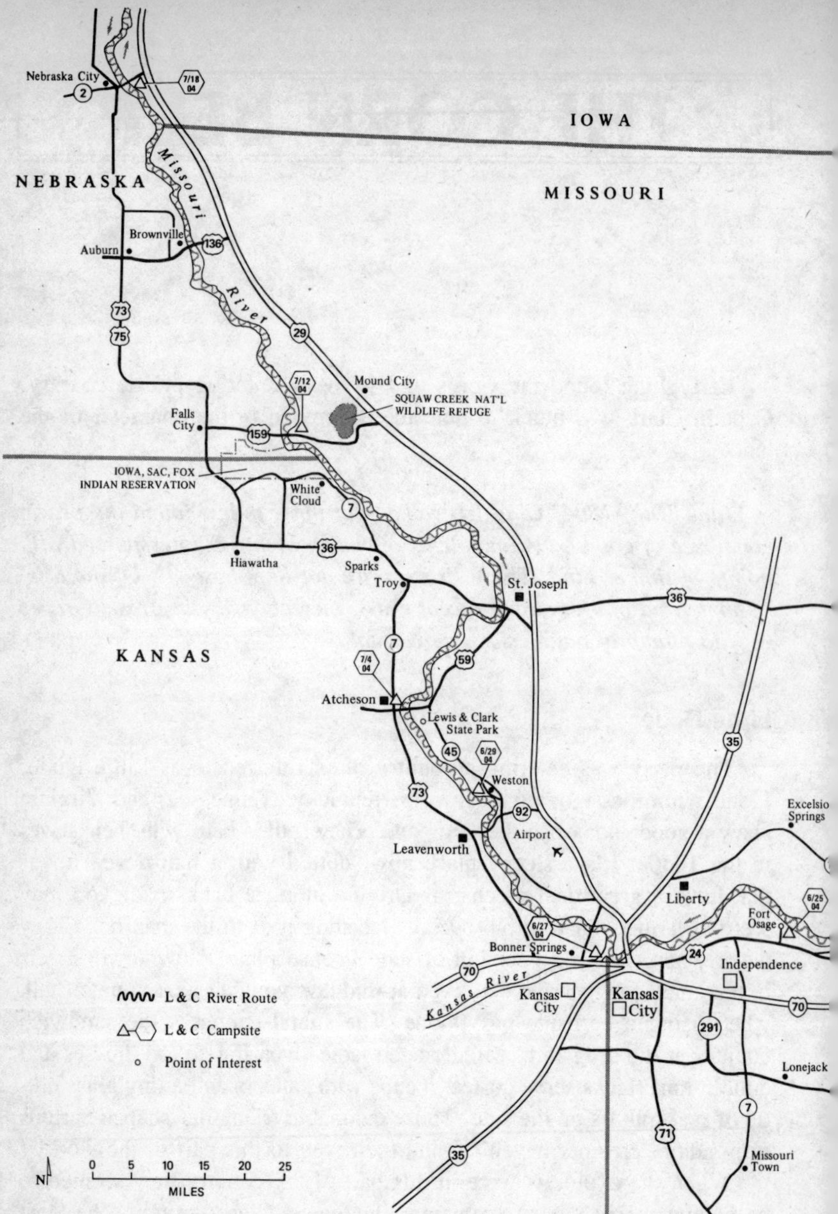

Nebraska City
2
7/18 04
IOWA
NEBRASKA
MISSOURI
Missouri
River
Brownville
Auburn
136
73
75
29
Mound City
7/12 04
SQUAW CREEK NAT'L WILDLIFE REFUGE
Falls City
159
IOWA, SAC, FOX INDIAN RESERVATION
White Cloud
7
36
Hiawatha
Sparks
Troy
St. Joseph
36
KANSAS
7
59
7/4 04
Atcheson
Lewis & Clark State Park
45
6/29 04
35
Weston
73
92
Excelsior Springs
Leavenworth
Airport
Liberty
6/25 04
6/27 04
Fort Osage
Bonner Springs
24
Independence
70
Kansas River
Kansas City
Kansas City
70
291
L & C River Route
L & C Campsite
Point of Interest
Lonejack
7
71
35
Missouri Town
N
0 5 10 15 20 25
MILES

IOWA

MISSOURI

THE CORN BELT

Clark tersely noted: "We had a find wind, but could not make use of it, our Mast being broke."

Boonville

The first town you come to in Little Dixie is Boonville, named after Daniel Morgan Boone, the frontiersman's son. For a while the town lay smack in the middle of U.S. 40, the main transcontinental highway, but with the construction of the nearby interstate it has reverted to its sleepy old self. It's easy to imagine white-hooded prairie schooners creaking up the hill, rolling past the red brick antebellum buildings that still grace the downtown. A few blocks away is the **Kemper Military School,** founded in 1844. Will Rogers was its most famous alumnus. The pride of the city is the Greek Revival structure known as **Thespian Hall.** Said to be the first theater building west of the Alleghenies, it plays host each summer to the Missouri River Festival of the Arts.

The first land battle of the Civil War was fought here in June of 1861. State troops under the command of the Confederate Colonel John S. Marmaduke were routed by federal forces under Captain Nathaniel Lyon. The victory was important in preserving Missouri for the Union.

Across the river, the town of New Franklin isn't really very new. It replaced the 1816 town of Franklin, which was destroyed by a flood in 1828. Franklin was the headquarters for the Boonslick farmers who pretty much ruled the territory in the early days. Another nearby town, *Rocheport,* is filled with antiques shops.

Arrow Rock

Northwest of Boonville is the picturebook town of Arrow Rock, situated on a bluff that Captain Clark noted would be a good site for a fort. The spot was named fully 80 years before the Corps of Discovery passed by. Historians believe the fur trader Dumont de Montigny named it Pierre a Fleche because the natives found flint here. More romantic souls say it was derived from a contest between a group of young warriors who assembled on a sand bar opposite the cliff to compete for the hand of the chief's daughter. The winner shot his arrow so far it lodged in the bluff. The bluff really isn't very prominent—it just happens to be the first one you come to when heading downriver.

Because of its proximity to the Boone's Lick salt works located directly across the river, Arrow Rock was, for a while, the starting point of the Santa Fe Trail. As noted earlier, the Boone's Lick Trace was the first wagon road west of the Mississippi. The first successful trading mission to Santa Fe, led by William Becknell, started here in 1821. The following year he took wagons over the route and returned with the first of the celebrated Missouri mules.

Today, you can still enjoy a meal, but not a drink in the tavern, built in 1834 to service those pioneers. The country store across the street is a combination grocery, antiques shop, and flea market. Dozens of stately buildings nestle among the elm trees. Natives sit on their front porches and watch the tourists walk by. Ask anyone if there is a room to rent and they'll not only know the answer, but will likely tell you where the hostess happens to be at that particular moment. This, incidentally, is MissourUH country, not MissourEE. The state officially admits to both pronunciations, but the "UH" is generally heard spoken by older people living in the west, while "EE" is more prevalent in the eastern part of the state.

Arrow Rock plays host to Missouri's other summer-stock theater company. Professional and semiprofessional actors team up to provide a lively summer repertoire at the **Lyceum Theatre,** once an old church.

Lexington

Driving west from Arrow Rock on State Highway 41, you soon really begin to feel like you're in the west. The sweet smell of freshly cut alfalfa permeates the air. Soybean crops alternate with corn, rye, and wheat. Occasionally you pass an apple or walnut orchard. The farms appear prosperous, judging from the well-kept homes. A sign out front identifies the owner as being a member of the Polled Herford Association. You will pass through **Marshall,** named for the Supreme Court justice, and about as typical a western farm town as you'll see. This is hog country. The settlers from Virginia, Tennessee, and Kentucky brought to Missouri one of its best-loved gourmet treats. Their recipes for curing and ageing ham, refined over the years, have resulted in justly famed Missouri country-cured ham. Smoked and aged to perfection, it is served with red-eye gravy and biscuits.

Twenty miles west of Marshall is the delightful river town of **Waverly,** population 900. The Corps of Discovery stopped here for a day to make oars out of the ash they found nearby.

June 17th, 1804 [Clark] George Drewyer our hunter came in with 2 Deer & a Bear, also a young Horse [which has] been in the Prarie a long time and is fat, I Suppose, he has been left by Some war party against the Osage, This is a Crossing place for the war parties against that nation from the Saukees [Sacs], Ayauways [Iowas], & Souix.

Twenty miles farther upstream is Lexington, site of two memorable events. In 1852 the side wheeler *Saluda,* carrying 250 Mormons, tried to get around a bend in the river but was repulsed by ice being swept downstream by a very fast current. The next day the captain, thwarted once again, demanded all steam possible. The crew responded by tying down the safety valve with the predictable result that the boiler exploded. The passengers were blown to bits and scattered over the river and its banks. Only 100 corpses were ever accounted for.

Nine years later another tragedy occurred. Colonel James A. Mulligan's Union forces established themselves on a nearby bluff with the aim of preventing Southern sympathizers from moving downstream to join the Rebel army. General Sterling Price, a hero in Missouri, laid siege to the hot and waterless Federals. He had his men construct breastworks of hemp bales soaked in water to withstand heated shot. One witness described the Federal soldiers as "dying from thirst, frenziedly wrestling for water in which the bleeding stumps of mangled limbs had been washed and drinking it with a horrid avidity." After 52 hours Mulligan surrendered the 3000 soldiers who were still alive. The battlefield is now a lovely state park overlooking the river. The red brick neocolonial **Anderson House,** which was used as a field hospital during those horrid hours, is now a museum, open to the public.

Present-day Lexington is a pleasant town of 5000 souls. Coal mining was once big here; 9000 people were employed in the mines, but now soybean farming and apple growing are the principal activities. The county courthouse is the oldest still in use west of the Mississippi. A fine old restaurant, the **Victory Rivertown Inn,** with its French Provincial decor, graces the main street.

Fort Osage

The next stop on our westward odyssey is Fort Osage. Instead of taking U.S. Highway 24, go down by the river and follow State Route 224 until it rejoins the highway. The reconstructed fort is on a well-signed road about 3 miles north of U.S. Highway 24. On June 23 the Corps of Discovery passed beneath the bluff, which Captain Clark noted "has many advantages for a fort and trading-house with the Indians." Two years after his return to St. Louis the Missouri Fur Company, with

Clark as one of the partners, did exactly that—they established Fort Osage.

It's appropriate to pause here and describe the fur trade as it existed two years after the return of Lewis and Clark. For a number of years trade with the Osages had been monopolized by the Chouteau family of St. Louis. Meriwether Lewis had been the guest of August Chouteau during much of the winter while the Corps of Discovery was encamped in Illinois. An outsider, however, one Manuel Lisa, had muscled his way into the picture. He was one of the most detested and feared men on the river. Swarthy and sinister, he seldom found it safe to turn his back on his own employees. Meriwether Lewis detested him, and a contemporary who knew him said: "Rascality sat on every feature of his dark complexioned Mexican face."

As it turned out, "rascality" was an essential attribute for anyone engaged in the fur trade, and the Chouteau family knew it. They reluctantly took in Lisa as a partner in the newly formed Missouri Fur Company. Meriwether Lewis's brother, Reuben, was part of the group, as was William Clark, though the propriety of his joining the venture was questioned by some because he was a public employee at the time. Other partners were later to become famous in the fur trade: Andrew Henry, Pierre Menard, and Sylvestre Labbadie, a well-educated member of an old St. Louis family.

Fort Osage was not a fort in the military sense, but simply a secure place to trade with the local tribes. The trading process was called factoring, the agent was the factor, and the fort the factory. Reflecting the French origins of this part of the country, the chief man at the fort was the *bouguise,* the employees were called *engages.* Trade goods consisted mainly of blankets, traps, guns and ammunition, tomahawks, knives, kettles, colored cloth, beads, silver ornaments, vermilion, and breech-cloth, for which buffalo hides, dressed and shaved deer skins, bear, beaver, raccoon, wolf, fox, badger, and muskrat furs were given in return. At one time some 5000 Osages settled about the fort, but in 1825 they signed a treaty giving the land to the newcomers. Two years later the post was abandoned.

You can poke around the blockhouses, living quarters, barracks, stockade, and trading post. During the summer native customs are demonstrated, and crafts are sold on the lawn nearby. A museum has excellent displays of the life and lore of the Little and Big Osage and Pawnee tribes. The surrounding park, which overlooks the river, is an excellent place for a picnic.

Missouri Town 1855

About 20 miles southwest of Fort Osage along State Highway 7 is Missouri Town 1855, where many of the state's old buildings have been carefully relocated. There are homes, barns, a stagecoach stop, blacksmith shop, livery stable, law office, and church. During the summer college students receive credit for living in the old buildings and spending a week cutting wood, planting and tilling gardens, making lye soap over an outdoor fire, cooking or baking in the hearth fireplaces, and caring for the livestock. The facility is open year-round.

Independence

Just west of Missouri Town is Independence, a thriving river port at a time when Kansas City didn't even exist. Daniel Morgan Boone was the first to visit the area. The city has a population of over 100,000, but the whole area has grown so that it's now mainly a suburb of Kansas City. Independence has done a superb job of presenting its history to tourists. At the visitors center located downtown, called the **Harry S Truman Historical District,** you can park your car and forget it for a few hours. The city runs a free shuttle bus that makes the rounds of all the historical sites. Next door to the visitors center is the **Marshal's Home Museum,** housing the jail where John Younger, Frank James, and William Quantrill were guests of the county for a spell. The furnished house displays many relics from the Civil War days. Nearby is the **Jackson County Courthouse,** where Harry S Truman began his political career as presiding judge of the county court. The bus stops at the **Truman Library,** where he and Bess are buried and where public galleries display mementos from his presidency. The lovely **Truman house** is open to the public, but you first must go to the tourist office and obtain a pass. They're free, but only a limited number are given out each day, and they are good only for the time specified. During the summer the day's allotment is often gone by midmorning.

As settlements pushed their way up the Missouri, Independence replaced Arrow Rock as the starting point for the Santa Fe Trail. In 1831 the Prophet Joseph Smith and his Mormon followers attempted to establish their city of Zion here. For a while they were accepted, but growing fear led a committee of townsfolk to draft a tract that became known as the Missouri Manifesto. It said in part:

> . . . *some two or three of these people made their appearance in Missouri; that they now numbered upwards of 1,200; that each succes-*

sive autumn and spring poured forth a new swarm of them into the country as if the places from which they came were flooding Missouri with the very dregs of their composition; that they were little above the condition of the blacks in regard to property and education; and that in addition to other causes of scandal and offense, they exercised a corrupting influence of the slaves.

With such sentiment, it's no wonder the Missouri slaveholders forced them out of Jackson County. **The Mormon Center,** which is open to the public, is designed to acquaint the visitor with the history and beliefs of the Reformed Church of Jesus Christ of Latter-day Saints.

Other interesting sights in Independence are the **Bingham-Waggoner Estate,** home of Missouri artist and politician George Caleb Bingham, Pioneer Spring, and Mount Washington Cemetery where mountain man Jim Bridger is buried.

Liberty/Excelsior Springs

The tragic Mormon story takes another turn across the river in the little town of Liberty. Here, in 1838, Governor Bogg ordered that "the Mormons must be treated as enemies and must be either exterminated or driven from the state, if necessary, for the public good." Seventeen Saints were killed in the riots that followed. You can visit the jail where Joseph Smith was held prisoner. Smith took his flock to Illinois where, in 1844, he was again imprisoned and subsequently murdered. It was a miracle, the Saints believe, that God brought forth Brigham Young to take his place and lead the flock to the Great Salt Lake where the followers of the Book of Mormon have prospered ever since.

Liberty holds another dubious claim to fame. It's where Jesse James robbed his first bank. He got $60,000 for his effort. The site is on the corner of the main plaza. Nearby is the **Clay County Museum,** housed in a former drugstore. The old hardwood shelves are crammed with turn-of-the-century nostalgia.

Ten miles northeast of Liberty is Excelsior Springs, home of the only European-type spa in the midwest. The delightful pastime of "taking the waters" is having a renaissance these days as people seek to unwind from their everyday stresses. The water here, heavy with calcium, sodium, iron, and manganese, is said to help cure such ailments as rheumatism, liver complaints, diseases of the kidneys and bladder, dyspepsia, and piles. Whatever the efficacy of this magical water, the treatment is just plain fun. Spas are located in the city-run **Hall of Waters,** built in

1937, and in the justly famous **Elms Hotel,** now approaching its hundredth year of operation. The 100-room hotel was rebuilt in 1912 and over the years has played host to the rich and famous, and in one case, the infamous—Al Capone. Harry Truman spent the night here, blissfully unaware that he had defeated Thomas E. Dewey in their race for the presidency. The refurbished gray limestone building is set in a 24-acre park and, in addition to the spa, has a fitness center, running track, and outdoor pool. A championship golf course is nearby.

Kansas City

June 26th–28th, 1804 [Clark] passed a bad Sand bar, where our tow rope broke twice, & with great exertions, we rowed round it and came to & camped, in the Point above the Kansas River I observed a great number of Parrot queets this evening. The Countrey about the mouth of this river is verry fine on each Side This river receves its name from a Nation which dwells on its banks those Indians are not verry noumerous at this time, reduced by war with their neighbours, &c, I am told they are a fierce & Warlike people, being badly Supplied with fire arms, become easily conquered by the Aiuway [Iowas] & Saukes [Sacs] who are bettter furnished with those materials of War, This Nations is now out in the Plains hunting the Buffalow.

The confluence of the Kansas River, locally called Kaw, and the Missouri River became the site of Westport, which is now the great metropolis of Kansas City. The bird Captain Clark was referring to was the now extinct Carolina paroquet.

Kansas City shares with Denver the honor of being the hub city of the trans-Mississippi West. With a metropolitan population of nearly one and a half million, it's large enough to support a whole host of activities from symphony to baseball. Kansas City lies a tiny bit to the east of the geographical center of the contiguous 48 states, so purists might argue it isn't even in the west, but if the west means a state of mind, then this is where it begins. Comparisons are often made between Missouri's two major cities. You hear statements like: "St. Louis is urbane and eastern; Kansas City is boisterous and western" and "St. Louis is catfish; we're steak." The city does everything it can to foster the western image: it's the home of the American Royal, the country's premier cattle show and rodeo. The baseball team took its name from that event. Football is represented by the Chiefs, and the civic booster organization is called the Sirloin and Saddle Club.

The city gained this image in the 1870s, when it was the terminus of the Texas cattle trail. With the coming of the railroad, the beef industry began to thrive. The American Hereford Association has its headquarters here, on a bluff overlooking the river bottom, where the stockyards and the American Royal facilities are located. There has been a marked change in this industry in the last 20 years, however. The stockyards no longer play a dominant role in the moving of beef to market. The great yards of Kansas City, St. Joseph, Omaha, and Sioux City do a fraction of their former business. Today, instead of taking the cattle to their food, meaning the open range, the food is brought to the cattle. Huge feedlots, often with a veterinarian on the staff, house thousands of head each. Diets are controlled with computer accuracy. It is therefore no longer necessary to bring cattle to the auction yards; buyers go to the feedlots to do their bidding.

The great metropolis of Kansas City grew from a rather shaky beginning. Francois Chouteau, son of Lewis's friend, built a trading post in 1821. The town, named Westport, was platted along the Big Blue River in 1833, but was later wiped out by a cholera epidemic. By the middle of the 1850s Kansas City was embroiled in the conflict between proslavery and antislavery forces over the settlement of the Kansas and Nebraska territories. Armed bands of Southern sympathizers launched forays into Kansas to stuff ballot boxes and terrorize Free State settlers.

In 1863 Confederate guerrilla leader Quantrill's band, which included Frank and Jesse James, sacked Lawrence, Kansas, prompting the Union General Thomas Ewing, Jr., to order anyone who could not prove his Union sympathies to vacate his land. The following year the largest battle west of the Mississippi, in terms of the number of combatants engaged, was fought near Westport. Urban growth has now completely engulfed the site of the battle; about the only memento today of the bloodiest day in Missouri's history is a bronze plaque, located in a traffic island on a busy four-lane thoroughfare.

The battle of Westport was fought on a cold October day in 1864. Trying to stem a series of Confederate setbacks (Sherman was at that time nearing Atlanta), General Sterling Price, the hero of Lexington, marched on the fledgling city with 10,000 cavalrymen. Arrayed against him was General Samuel R. Curtis and the Army of the Border and the Kansas militia with 12,000 men. The Confederates were well fortified, but a farmer showed Curtis a way up a defile that pierced a gap in the enemy lines. This was the beginning of the route that later saw a pistol duel on horseback between Capt. Curtis Johnson of Kansas and Col. James H. McGhee of Arkansas. McGhee fired first, but the wounded

Johnson replied with a fatal shot through the heart. The now leaderless Rebels fled from the field. About a thousand men were killed, wounded, or captured that day, about equally divided between the two sides. It was to prove to be General Price's swan song. To learn more about these events, drive southeast on U.S. Highway 50 some 30 miles to the town of **Lonejack,** where **the Civil War Museum** of Jackson County is located.

The more interesting things to see and do in Kansas City are near the site of the battle. General Curtis's troops were camped at what is now the Country Club Plaza, said to be America's oldest shopping mall. This center, with its Spanish Colonial architecture, features brick roads, statues, and fountains. Nearby is **Westport Square,** a restored historic district filled with shops, restaurants, and nightlife. The oldest building in Kansas City is here; built in 1837, it was once owned by Daniel Boone's grandson, and has for a long time been one of the favorite watering holes in the city. A few blocks east of Country Club Plaza is the **Nelson Gallery of Art** and **Atkins Museum,** which houses a collection of paintings, sculpture, and oriental art that rivals anything St. Louis has to offer. Two first-class hotels are located nearby.

The downtown area is about 3 miles north of Country Club Plaza. Here you'll find numerous fine stores and the spanking new convention center. There are several first-class hotels in the downtown area including the famous old **Muehlback,** now restored as the Radisson Muehlbach, which boasts that it has hosted every president since Woodrow Wilson. A mile south is the **Crown Center,** a showplace built by Hallmark Cards. The center features two first-class hotels, a shopping mall, and an outdoor plaza, where concerts are held during good weather. Hallmark Cards offers tours of a nearby manufacturing area. To the casual observer, the center doesn't seem a huge success; everybody goes home after working hours, and many shop spaces remain unrented. Across the street is the cavernous **Union Station,** once one of the most handsome railroad depots in the United States. Amtrak still uses the station, but the heating bills were so high they built an air-supported bubble inside the room. Today's travelers wait in 20th-century ugliness. Another nearby attraction is the **Liberty Memorial,** America's only museum devoted solely to the First World War. The **Torch of Liberty,** a 200-foot tower, has an elevator leading to a viewing platform.

You get a feeling for the upbeat go-gettedness of the Kansas City people when you drive by the **Truman Sport Complex.** Not one, but two separate stadiums were built at the same time, one for football, the other, baseball. Adjacent to the complex, near the junction of Interstates

70 and 35, the city operates a visitors center. A large staff will give you advice on where to stay and where to eat. The most often asked questions, according to the lady behind the counter, are where to find the Golden Ox and Arthur Bryant's. The **Golden Ox** is a huge steak house near the stockyards. **Arthur Bryant's** is a more modest, cafeteria-style barbequed ribs place, which *New Yorker* writer Calvin Trillin unabashedly calls the finest restaurant in the country. He's right. The whole town mourned Bryant's death. A local newspaper ran an editorial cartoon showing Arthur at the Pearly Gates. St. Peter plaintively asks: "Did you bring the sauce?"

South of town is the huge **Swope Park,** where summer plays and concerts are given in the outdoor **Starlight Theatre.** A fine zoo is located nearby. The city operates an historical museum in the former estate of lumber baron Robert Long. During the summer you can take an excursion aboard a barge pushed by the former tow boat *Westport*. The dock is located at the foot of Grand Avenue.

The **Agricultural Hall of Fame** is in **Bonner Springs,** Kansas, 18 miles west of Kansas City near Interstate 70. This museum has the largest collection of farm implements in the country, including steam threshing engines, a replica of an 1831 McCormick reaper, sodbuster plows, old farm wagons, and trucks.

Weston

The explorers were five weeks out of St. Charles when they left the Kansas River. From now until their winter camp in North Dakota they would be heading primarily north, not west. Ironically, they had their only trouble with whiskey 30 miles upstream, where they passed the site of the oldest still-operating distillery in America.

June 29th, 1804 [Clark] A Court Martiall will Set this day John Collins Charged "with getting drunk on his post this Morning out of whiskey put under his charge as a Sentinal. Hugh Hall Charged with takeing whiskey out of a Keg this morning which whiskey was stored on the Bank and under the Charge of the Guard.

Collins was given 100 lashes on his bare back, Hall 50. If either had been around 50 years later, they could have drunk all the whiskey their pocketbooks could buy. Ben Holladay founded a distillery, but soon sold it to the McCormick family and went off to do other things, like run a stagecoach line and found the Pony Express. Limestone springs were the reason for selecting this site. The product is as tasty as anything Kentucky has to offer. Tours of the ancient stills are offered daily, but, unlike a winery, federal law prohibits on-site sales. You have to go into town to sample the product.

The tiny town of Weston, population 1500, has nearly 200 pre-Civil War homes and businesses in a 22-block area. Stern-wheel steamboats used to tie up at the foot of Main Street, but the capricious river moved, and the town is now land-locked. Down near the tracks is the musty-smelling Burleigh tobacco warehouse, the only one this side of the Mississippi. The town boasts a number of antiques shops, two wineries, and one of the finest restaurants in Missouri, the **America Bowman Keeping Room.**

Weston lies in a section of Missouri called the Platte Purchase. Originally, the state boundary went due north from the mouth of the Kansas River. Everything west of that line was Indian territory ceded to the Sac, Fox, and Iowas in the Prairie Du Chien Treaty of 1830. Settlers, however, felt the boundary should be the Missouri River, thus ensuring unrestricted use of that natural highway. Captain Clark negotiated the purchase in 1836. In return for their land, the tribes received $7500 in cash, the services of a blacksmith, schoolmaster, and ferry boat, a year's rations, and 400 sections of land in Kansas. The treaty set the pattern that, 50 years later, would prompt the Ogalla Sioux Chief Red Cloud to comment: "They made us many promises, more than I can remember, but they never kept but one; they promised to take our land and they took it."

Leavenworth

Directly across the river from Weston is Leavenworth, Kansas, a name that, for most of us, evokes the same bad thoughts as San Quentin, Joliette, and Sing Sing. Leavenworth is, as they say, "the big house," but it is also a pleasant community that is home to a very historic army base. Col. Henry H. Leavenworth established a fort here in 1827 to keep the peace in the west and to guard the Santa Fe Trail. Of the several dozen army forts we'll visit on the upper Missouri and the Columbia, it's the only one still active as a base for military operations. The army established its General Command College here in 1881. It has since seen the country's most famous generals pass through its halls, including such Second World War heroes as George C. Marshall, Omar Bradley, Hap Arnold, Matt Ridgeway, Mark Clark, and Dwight D. Eisenhower.

The fort is the site of several of the most historic buildings in Kansas, including the **Rookery,** built in 1832 as the temporary home of the first territorial governor, the lovely **Memorial Chapel,** and the **U.S. Disciplinary Barracks,** built in 1874. The **Fort Leavenworth museum** has a large collection of horse-drawn vehicles and depicts the Army's role in the opening of the west by means of dioramas, displays, and audio-visual presentations. A lovely park overlooks the river. The cannon on display here were made in Paris in 1774.

By 1858 the fort was the major staging area for army supply trains heading west. The firm that won the contract for this lucrative freighting business was Russell, Majors and Waddell, who, at one time, had 3500 wagons, and employed 4000 men to drive an incredible 40,000 oxen.

Thirty-five thousand people live in Leavenworth. The downtown shopping area is graced with awning-covered sidewalks. The drive north to Atcheson is along State Route 7, which wanders through some nice wheat fields.

Atcheson

In 1872 the citizens of this quiet farm town floated a bond issue to build a railroad from here to Topeka, Kansas. The line prospered well enough to extend its tracks to Santa Fe, and thus was born the Atcheson, Topeka and Santa Fe, one of the largest railroads in the country. The railroad put an end to the fortunes of D. A. Butterfield's stagecoach operations, which were headquartered here. At the height of the firm's prosperity, Butterfield employed 1500 mule-skinners and bull-whackers to drive his immense fleet of wagons and stagecoaches. Ben Holladay, after selling his distillery in Weston, bought the bankrupt Russell, Majors and Waddell Freighting Co. and moved it here from Leavenworth.

The town was named after a proslavery Missouri senator who wrote a paper titled *Squatter Soverign,* and whose favorite pastime was to tar and feather Free-Staters. David Atcheson, who was president pro tem of the Senate, held the office of president of the United States for one day when Zachary Taylor refused to take the oath of office on a Sunday. Atcheson has a number of buildings on the National Register of Historic Places, including the home of aviatrix Amelia Earhart, which is now a museum. A lovely park is located on the river where Independence Creek comes in from the northwest. Lewis and Clark gave the creek its name.

July 4th, 1804 [Clark] We came to and camped in the lower edge of a Plain above the mouth of a Creek [which] we call Creek Indepen-

dence as we approached this place the Prairie had a most butifull appearance Hills & Valies interspsd with Copses of Timber gave a pleasing deversity of the Senery. we closed the day by Descharge from our bow piece, an extra Gill of whiskey.

Across the river, on the Missouri side, is **Lewis and Clark State Park,** which includes a lake Captain Clark named Gosling Lake. Sergeant Floyd, who was to die a few weeks later, described this country as "open and butifulley Divided with Hills and valies all presenting themselves." The park is a great place for a picnic.

St. Joseph

The days were getting hot as the party neared present-day St. Joseph, and the Corps of Discovery was still having its share of problems.

July 4th, 1804 [Clark] Jos. Fields got bit, by a snake, which was quickly doctored with bark by Cap Lewis. July 7th. one man verry sick, Struck with the Sun. Capt. Lewis bled him and gave Niter which revived him much.

Lewis was following the accepted medical practice of the day. The bark was a mixture of tree bark and gunpowder, placed over the wound and ignited, providing a cauterizing effect. The niter treatment was a dose of saltpeter which, when swallowed, increases urine discharge and induces sweating. The journals often comment on the dubious practice of blood-letting. A few days later Clark mused:

July 20th, 1804 [Clark] It is worthey of observation to mention that our Party has been much healthier on the Voyage than parties of the same number is in any other Situation. Tumers have been troublesome to them all.

With a population of 80,000, St. Joseph is Missouri's third largest city and one of its oldest. Joseph Robidoux (pronounced Roobidoo), a fur trader, selected the bluffs above the river as a perfect site for a fort and trading post, which he named after his patron saint. Robidoux's role in the settling of the west was significant. He had previously established a fort at Council Bluffs in competition with Manuel Lisa. The two fur merchants formed a pact that neither would take advantage of the other, but Lisa, the scoundrel that he was, secretly prepared to do just that with a band of Pawnee. To appear innocent of duplicity and prevent Robidoux from himself trading with the Pawnee, he paid a call on his neighbor. Complaining of illness, Robidoux, so legend has it, asked Lisa to fetch a bottle of champagne from the cellar. He thereupon let the trapdoor fall, rolled a cask upon it, and with mocking words, left his opponent imprisoned while he alone made his trades with the Pawnee.

The city played a huge role in the opening of the west, though no townsman would have thought so that day in 1844 when an itinerant Swiss wheelwright by the name of James W. Marshall disembarked from a riverboat. Marshall went west, accompanying a wagon train of 800 pioneers. A few years later, while building a sawmill for his employer, John Sutter, he found gold on the American River near Sacramento, and the rush was on.

Nowhere on your trek toward the Pacific do you get a better feel for how the pioneers must have felt as they started west than you do by driving to the bluff northwest of downtown St. Joe and looking out over the river. There, stretching out as far as your eye can see, is "the west," looking much as it did in the days of the prairie schooner. Even today, there is no industry and few farmhouses to spoil the illusion. You, too, could be a forty-niner, heading west to seek your fortune. Nearby, on the bluff, is **Robidoux Row,** which dates from the 1840s and is said to be the first apartment building west of the Mississippi. The restored buildings are open on weekends.

St. Joseph, as most schoolchildren know, was the eastern terminus of the Pony Express. Ben Holladay, of stagecoach fame, was one of the backers of the enterprise. On April 3, 1860, little Jonnie Frey, one of 200 young, small, and wiry riders chosen for their stamina, mounted his horse carrying a buckskin pouch with 85 pieces of mail and headed for the river ferry. His was the first lap of a 1975-mile trip to Sacramento, California. The trip took 7 days, 19 hours. The stable where Jonnie's horse was kept has been restored and made into a museum, recreating the romance of that great adventure. What most of us forget is that this bold experiment only lasted 19 months and was a financial failure. The stringing of telegraph wires across prairie and desert spelled its doom.

The Pony Express proved to be St. Joe's swan song in its role as a transcontinental jumping-off point. By the mid-1860s the country was ready to bridge the continent with rails of steel; the only question was where the eastern terminus should be located. St. Joe, with its long history as an outfitting point and its easy access to the Mississippi via the well-built and profitable Hannibal and St. Joseph Railroad, was the most likely candidate. It no doubt would have been selected had it not been for the Civil War and the uncertainty of where the Missourians' sympathies lay. The fact that there were 2000 slaves living in this area at the time made Abraham Lincoln's decision easier: The transcontinental railway would begin in a state he could count on. Thus, Iowa won the battle. Council Bluffs became the starting point for America's first railroad to the Pacific and St. Joe more or less faded into the backwater of western expansion.

Though its pioneer role was over, St. Joseph continued to claim a place in U.S. history. For a while the city was the end of the Texas cattle trail, and slaughterhouses were built that continued to be the backbone of that industry until the 1970s. Jesse James came to live here, incognito, and was murdered by his former friend and accomplice Bob Ford, who was assisted by his brother Charles. The two wanted the $10,000 bounty on Jesse's head. The murderers hardly profited from their crime. They managed to get a hanging sentence commuted by the governor, but, having moved to Colorado, they lived in debauchery until Charles committed suicide and Robert was shot in a dance-hall brawl. The house where Jesse's murder took place was moved from its former location to a site next door to the Patee House Museum. For a fee you can go inside and see the bullet hole, up on the wall near the picture Jesse was hanging at the time the dastardly deed was done.

The **Patee House,** built in 1858 as a hotel, served as headquarters for the Pony Express. In the Western Americana collection is an old wood-burning locomotive and reconstructions of old dentists' offices, drugstores, and the like. St. Joseph also boasts the **Albrecht Art Museum,** with its collection of 18th-, 19th-, and 20th-century American art, and the **St. Joseph Museum,** housed in a 19th-century Gothic sandstone mansion overlooking the river. The house is lovely, with genuine Tiffany glass windows. The collection includes exhibits on the Native American, natural history, western expansion, Pony Express, Civil War, and Victorian periods.

St. Joe's downtown has many magnificent brick buildings and a new pedestrian mall. You can dine on Mexican food in a castlelike building dating from 1890. Much of the commerce, however, has fled to the shopping center out near the interstate, where the motels are also located.

Iowa, Sac, Fox Reservation

To avoid the monotonous interstate, it's best to cross the river at St. Joseph and proceed north on the Kansas side. U.S. Highway 36 heads west, past the lovely little apple-growing town of *Troy*. Take the time to visit the old brick copper-domed courthouse. Inside, the recorder's office has a turn-of-the-century feel, yet outside is a modern sculpture of an Indian head, carved out of a huge redwood log. Sitting in camp, far upstream from here, the explorers had time to reflect on this country:

> *August 21st, 1805 [Biddle] On the south the hills continue close to the river from the ancient village of the Kansas up to Council Bluff, 50 miles beyond the Platte, forming high prairie lands. On both sides the lands are good; and perhaps this distance from the Osage to the Platte may be recommended as among the best districts on the Missouri for the purpose of settlers.*

North of Troy is the site of the Iowa, Sac, and Fox Mission, an Indian school established by the Presbyterians in 1837. Like many we will see going west, the mission was a failure. The schoolmaster wrote that the Iowa "were a wild, warlike, roving people. They depended mainly on the hunt for subsistence, with war as a chief pastime. The women did what farming was done while the men danced, raided neighboring tribes or consumed alcohol." The missionaries tried to teach the youngsters to speak English, how to farm, and, of course, how to worship the white man's God. With the encouragement of their parents, the children simply refused to go to school.

The closing of the mission in 1860 did not have an appreciable effect on the government's search for a solution to "the Indian problem." A few years later the idea in vogue was to set up boarding schools off the reservation (patterned after the one in Carlisle, Pennsylvania), where the Indian children could be made to speak only English, and where military discipline would force them out of their slovenly ways. The graduates, having lost their tribal culture, could no longer even converse with their parents and were ostracized by the rest of the tribe. No one else wanted them either.

White Cloud

You can see some nice river country by turning north on State Highway 7 and driving to the tiny town of White Cloud, Kansas, which was named after a chief of the Iowa Nation. The one-block-long main street with its red-brick buildings is an inviting place to stop for a snack. You have a beautiful view of the river and the surrounding countryside from the bluff north of town. On a reasonably clear day you can see four states: Kansas, Missouri, Nebraska, and Iowa.

Lewis and Clark camped nearby.

> *July 10th, 11th, 1804 [Clark] We camped on the S.S. [starboard side] opposite a yellow Clay Clift, Capt Lewis Killed to young Gees or Goslings. The men of the party getting better, but much fatigued. The bottom is verry extensive on the S.S. and thickly intersperced with Vines. The High Land approaches near the river and well timbered next to the river, back of those hills the Plains commence. Seven Deer killed to day, Drewyer killed six of them, made some Luner observations this evening.*

To continue upriver without having to retrace your steps, head west out of White Cloud on the unnumbered county road. It's along about here that you may spot your first pickup sporting rifles mounted on a rack behind the driver—the sure sign that you've finally arrived in the west. This is Chevy and Ford country; maybe one vehicle in ten is foreign-made, and the pickups outnumber the sedans by a sizable margin.

The drive over the undulating prairie is lovely. The country looks exactly like Kansas wheat fields are supposed to look. It's the black-and-white opening scene from *The Wizard of Oz*. After driving about a dozen miles you come to a *T* where you turn right. The road goes north for a little bit and then heads west again, where you soon join U.S. Highway 73 heading north. Just after crossing into Nebraska, you pass the Sac and Fox Indian Reservation. A sign indicates that part of it is set aside for half-breeds, mostly those descendants of French Canadian fur trappers who frequented these parts, even before the Corps of Discovery passed through. My reaction—a reservation for half-breeds? There's a problem I didn't even know existed.

Squaw Creek

Across the river, tucked in the northwest corner of Missouri, is the Squaw Creek National Wildlife Refuge, a major stopover for migrating waterfowl. Over 200 bald eagles nest here during the winter. Thirty-three kinds of mammals, 35 species of reptiles and amphibians, and nearly 300 species of birds have been seen in the marshes, lakes, cordgrass prairies, willow thickets, and cottonwood forests. Footpaths provide access to several viewing towers. Overlooking the refuge from the east are the loess bluffs, a rare geologic formation of wind-deposited soil found only here and at a site in China. On top of the bluffs are some of the last remnants of the once vast native prairie that dominated the area prior to the settling of the west.

Brownville

Another fine old river town is Brownville, Nebraska, whose present population of 250 is half what it was in its heyday. Among the many 1860s-style buildings is the refurbished **Carson House,** now open to the public, the **Gates House,** said to be the first brick building in Nebraska, and the **Beehive,** the state's oldest apartment building. Nebraska Wesleyan University sponsors a repertory theater here during the summer. Another summer attraction is the *Belle of Brownville,* a steam ferry that makes afternoon and evening excursions on the river. Permanently docked just down river is the *Meriwether Lewis,* an old Corps of Engineers' dredge that has been restored and converted into a free museum where you learn about cutterhead dredges, dust pan dredges, snag boats, and a bit about today's river traffic.

Nebraska City

This pleasant town of 7000 Nebraskans is the oldest in the state, built on the site of a short-lived fur trading post known as Fort Atkinson. The town's most illustrious citizen, J. Sterling Morton, took up a home site by squatter's right in 1855, built a frame house, and started planting trees. A newspaperman by profession, he made the planting of trees his life's work, first as president of the State Board of Agriculture and then through his influence as Secretary of Agriculture under Grover Cleveland. To him we owe Arbor Day, now celebrated on his birthday, April 22. The trees he and his wife planted are now part of an arboretum, managed by the Nebraska Game and Parks Commission. A walk through the pine grove, the American Chestnut stand, and the Italian terraced gardens is an enchanting experience. Morton's eldest son, Joy, made a fortune in the salt business and converted his parents' rather modest home into a 52-room neocolonial mansion that is now open to the public. Furnishings are largely in Victorian and Empire styles.

Nearby, an old house, connected to a cave by a passageway, was a major station of the Underground Railway, the route used by blacks fleeing

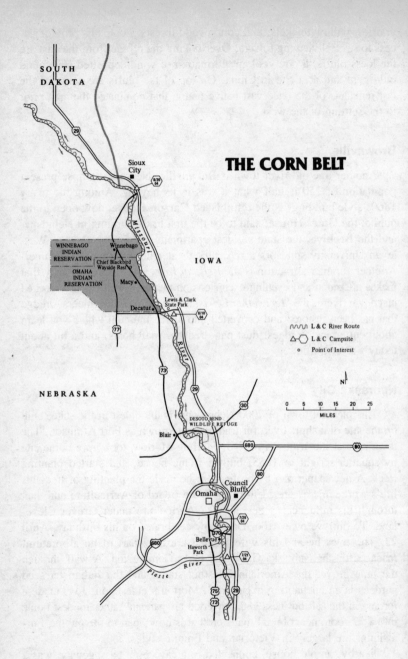

from Missouri and Kansas into Canada. For a century it's been known as the **John Brown Cave,** though historians disagree whether the future martyr at Harper's Ferry actually lived here. The site is now a privately operated museum.

Missouri/Platte Confluence

Most travelers following this part of the Missouri River take Interstate 29. Huge green-and-white billboards point the way: "Junction I 80, 36 mi, Council Bluffs, 42 mi, Sioux City, 165 mi." None of these landmarks existed at the time of Lewis and Clark, but landmarks, nevertheless, did exist. Furthermore, they were well known to the Corps of Discovery. Having no mountains or prominent rocks on the great American prairie to focus on, cartographers had only river junctions to put on their maps. A billboard along the Missouri in 1804 might well have read: "Platte River, 22 mi, Little Sioux River, 155 mi, Niobrara River, 224 mi."

July 21st, 1804 [Clark] Set out early under a gentle breeze proceeded on verry well. at 7 oClock the wind luled and its Commns'd raining, arrived at the lower Mouth of the Great River Platt. This Great river being much more rapid than the Missourie comes with great velosity roleing its Sands into the Missouri, filling up is Bead & compelling it to incroach on the North Shore. I am told by one of our Party who wintered two winter on this river, that "it is much wider above, and cannot be navagated with Boats or Perogues.

To hordes of pioneers who were to follow the great Platte River in the years to come, it certainly was not navigable, for according to the old saw, "It is a mile wide and an inch deep." The water was terrible, too: "It is water you have to chew" or "It was good drinking if you threw it out and filled the cup with whiskey." Nevertheless, in a few years it became the major route to the west. Zebulon M. Pike charted the way while Lewis and Clark were still on their way home from the Pacific.

Two factors accounted for the popularity of the route. As we will see, Lewis and Clark had some touchy dealings with the Arikaras and Sioux farther upriver, and subsequent travelers, starting in 1811 with the Wilson Price Hunt party of Astorians, simply decided to finesse the problem by going another way. Second, the Platte route provided a much easier passage over the Continental Divide, as Jedediah Smith found out in 1824 when he discovered Wyoming's South Pass. The Platte became the route of the Oregon, Overland, and Mormon trails, and the first transcontinental railway followed its banks. So much settlement was diverted west here that it wasn't until the building of the Northern Pacific Railway in 1877 that any appreciable immigration took place along the route pioneered by Lewis and Clark.

Bellevue

The first fur trading post in Nebraska was probably Fort Bellevue, built by Manuel Lisa and named for the fine view of the confluence of the Platte and the Missouri. All trace of the fort is gone, yet we do have Karl Bodmer's fine watercolor at the art gallery in Omaha to illustrate what it must have looked like. A number of old downtown Bellevue buildings have been refurbished, including the first church in the state, an old bank, a railway depot, and an 1830s' settler's log cabin. The **Sarpy County Historical Museum** is open on weekdays. Today, Bellevue is the home of **Offutt Air Force Base,** headquarters of the Strategic Air Command. The base has a museum, open to the public, with exhibits of fighters, bombers, and rockets, all of which will delight any pilot or airplane buff.

To get a good view of the Platte/Missouri confluence, drive north from the SAC base, and turn east on Highway 370. Just before the highway crosses the Missouri on a toll bridge you come to Haworth Park, where an excursion boat makes a 2-hour river trip every afternoon during the summer.

Omaha

Omaha rather than Bellevue became the principal city in the area, because Douglas county was quick to float bonds for the building of the first transcontinental railway. Things got off to a rather inauspicious start. Development began in 1852 when land sharks, speculators, and settlers started to congregate across the river in Council Bluffs, awaiting a treaty between the government and the Omaha Nation that would open Nebraska for settlement.

The first settlers were a hardy lot, mostly farmers and common laborers who organized themselves into what they called the Claim Club. Its purpose was to thwart government laws by trying to get 320 rather than 160 acres for each settler. In order to forestall the requirement that each man improve his property, they built a house on wheels that could be moved from one claim to another. Gunfights in the streets were common during the early years, and, so the story goes, two horse thieves were once lynched with a single rope. A noose was tied on each end and then thrown over a tree limb.

Omaha's 300,000 population makes it Nebraska's largest city, and it is the last real metropolitan center on your Lewis and Clark trek until you reach Portland, Oregon. Its revitalized downtown gives the city a cosmopolitan, vibrant, buoyant feel. A grassy mall, complete with flowing brook and modern sculpture, runs between the main business area and the river. A section of the warehouse district has been renovated with fancy shops, bars, and restaurants featuring Greek, Spanish, Mexican, Creole, and French cuisine. Flowering vines climb up old ironwork awnings; sidewalks are graced with geranium-filled planter boxes, the streets are paved with red brick.

Omaha's showplace is the **Joslyn Art Museum,** which boasts an art collection rivaling that of both Kansas City and St. Louis. The collection includes 16th-century masterpieces and works from such varied artists as Rembrandt, Titian, Van Dyke, Picasso, and Mary Cassatt. The gallery, however, is most noted for its collection of the drawings, watercolors, and lithographs of Swiss artist Karl Bodmer.

Bodmer was in the employ of a Prussian naturalist with an imposing name; Alexander Phillip Maximilian, Prince of Wied and Neuwied. In 1833 he visited with Captain Clark at St. Louis prior to sailing upstream on the American Fur Company's new steamboat, *Yellow Stone*. The voyage was memorable because Max, the scientist, and Karl, the artist, produced the most accurate description of the upper Missouri since Lewis and Clark. The journals and drawings were published under the title *Reise in das Innere Nord-America in den Jahren 1832 bis 1834*. The steamboat *Yellow Stone* made better time than the Corps of Discovery's keelboat did 30 years earlier—3 weeks versus 9 weeks to Omaha—but the voyage was frequently delayed because of the need to take on more wood for the boilers and because, at that time, the river was still too tricky to attempt to sail after dark. Fortunately, this slow pace gave Bodmer time to paint, and so the Joslyn collection includes a drawing of the *Yellow Stone*.

Maximilian and Bodmer spent a week at Fort Pierre (now Pierre, South Dakota), during which time 7000 buffalo hides and, reflecting on the gastronomical tastes of the day, 10,000 pounds of buffalo tongue were loaded onto the steamboat, which then returned to St. Louis. The party then sailed upriver on another steamer, the *Assiniboine,* arriving at Fort Union on the mouth of the Yellowstone River in late June. This was the head of navigation for steamboats in 1833, so, after a couple of weeks' delay, they traveled the last 500 miles to Fort McKenzie by keelboat. Fort McKenzie was at the mouth of Marias River, in the heart of Blackfeet country. Maximilian was the first scientist to study this ferocious tribe; he and Bodmer witnessed them in battle with a band of Cree and Assiniboines. (See page 114.)

On the return trip the party spent the winter near Lewis and Clark's Fort Mandan, where Bodmer whiled away the time painting the chiefs in their royal regalia. Most of his subjects were dead in a few years, victims of smallpox and other white man's diseases, so Bodmer's paintings truly reflect *Views of a Vanishing Frontier,* the title of a recent book describing the adventure.

The Joslyn is more than an art museum; it houses the auditorium where the symphony plays and theater companies perform. Downstairs is a very fine history museum, tracing life on the Nebraska plains from the prehistoric tribes through settlement by the whites. Lewis and Clark buffs will enjoy the early maps of the Missouri River because they illustrate some of what the captains knew before they set out from St. Louis.

Omaha has two other fine museums, both relating to the coming of the railway and its impact on the settling of the west. One is on the ground floor of **Union Pacific's headquarters building.** There are models of locomotives, a collection of guns, and, most interestingly, the tablewear and furniture from the railway car that carried Abraham Lincoln's body back to Springfield, Illinois. The other museum is located in the old **Union Station,** built at the height of the Art Deco movement of the 1930s. A collection of old telephone equipment and railway signaling devices shows early railroading practices, and numerous photographs illustrate the glamour of traveling by train during its heyday just prior to World War II. The lower floor, at track level, is still being developed. When completed it will be a tableaux of Omaha's development.

Betting on horses is legal in Nebraska. The local track, **Ak-Sar-Ben** (Nebraska spelled backward), draws much of its crowd from Kansas City, perhaps a quid pro quo for all the Omahaians that make the 190-mile drive south to watch baseball and football. Like Kansas City's American Royal, Ak-Sar-Ben is a civic booster organization that also sponsors stock shows, a rodeo, and a fancy-dress ball in the winter.

Gerald Ford was born in Omaha. His family house is open to visitors. The **Great Plains Black Museum** preserves the history of Black Americans' contribution to the settling of the west. The **Henry Doorly**

Zoo is open from April through October. You can also visit the campus of **Creighton University,** founded in 1878.

Council Bluffs

Council Bluffs owes its name, if not its existence, to the voyage of Lewis and Clark. It was here the party had its long awaited powwow with the Otos and Missouris. Jefferson's instructions to Meriwether Lewis make it clear that he wanted them to act as ambassadors.

June 20th, 1803 [Jefferson] In all your intercourse with the natives treat them in the most friendly & conciliatory manner which their own conduct will admit; allay all jealousies as to the object of your journey; satisfy them of its innocence; make them acquainted with the position, extent, character, peaceable & commercial dispositions of the U.S.; of our wish to be neighborly, friendly & useful to them, & of our disposition to a commercial intercourse with them . . .

The council was held on August 3 and was the precursor of many to follow. They met under an awning of sailcloth. The rites included a parade, speeches by one of the captains and the chiefs, smoking the pipe, awarding of medals and certificates to the Indians, exchanging of gifts, and, to awe the natives, a technological display involving such items as the party's air gun, magnet, spyglass, compass, and watch. The outstanding characteristic of these councils was the sincere recognition accorded by Lewis and Clark to the dignity of the Native Americans.

Clark found the country agreeable:

August 1st, 1804 [Clark] The Prarie which is situated below our Camp is above the high water leavel and rich covered with Grass from 5 to 8 feet high interspersed with copse of Hazel, Plumbs, Currents (like those of the U.S.) Rasberries & Grapes of Dift. Kinds. The Situation of our last Camp Councile Bluff appears to be a very proper place for a Tradeing establishment and fortification.

The city of Council Bluffs, or CB as the natives often call it, erected a monument to the expedition. Drive north through **Big Lake Park,** go up the hill on a gravel road, and turn left on the signed road leading to the most prominent bluff in the area. The monument is in sad repair, but the view out over the river and the skyline of Omaha is lovely.

The first steamboats arrived in Council Bluffs in 1819, a scant 15

Sioux Encamped on the Upper Missouri: George Catlin. Catlin was the first American artist to follow in Lewis and Clark's footsteps. This scene was painted in 1832. National Museum of American Art, Smithsonian Institution, gift of Mrs. Joseph Harrison, Jr.

years after Lewis and Clark had labored so hard to get here. They brought an army of 1000 soldiers under the command of General Henry W. Atkinson, who was on his way to the mouth of the Yellowstone, where he was directed to check British advances into the area and give a show of force to the local tribes. The so-called Yellowstone Expedition was the brainchild of John C. Calhoun, Secretary of War in the Monroe administration. But it was a fiasco; Council Bluffs was as far as they got. A monument marking the location of Fort Atkinson is along the river about 20 miles north of town.

Two monuments in Council Bluffs relate to the coming of the railroad. A spire commemorates Lincoln's visit, when he made his decision to make this the terminus of the transcontinental line. Construction began in 1865, and soon after a number of dignitaries, including General William T. Sherman, riding on flat cars with nail kegs for seats, took the first train west. Another monument, a 50-foot-high replica of the golden spike, celebrates the completion of the railroad four years later. In spite of the significance of the ceremonies at Promontory Point, Utah, it wasn't truly a transcontinental railway until a bridge was built across the Missouri between Council Bluffs and Omaha, which didn't occur until 1872. The Civil War General Grenville M. Dodge was the surveyor for the line, and because of his friendship with Lincoln he became its principal advocate. His house, a red brick beauty, is the showpiece of the town. It's now a museum, maintained with its original furnishings and artworks. The old **Pottawattamie County Jail** is another interesting place to visit. Designed in a "Lazy Susan" fashion, all the prisoners could be tended to by a single jailkeeper.

In the 1840s Council Bluffs was a Mormon town with a population of over 8000. When Brigham Young took his flock to Salt Lake City less than 10% of the citizens stayed behind, many of whom were the followers of Father De Smet, who established a mission here. His name will crop up many times on our westward journey.

DeSoto Bend National Wildlife Refuge

In the days before the Corps of Engineers started tinkering with the Missouri, the river often formed ox-bow lakes. The one at DeSoto Bend is so lovely it's worth considering how it was formed. Most rivers with a shallow gradient tend to meander, like the river Meander in Phrygia, which, according to Bullfinch, "returns on itself and flows now onward, now backward, on its course to the sea." When the flowing Missouri hit a slight bend, perhaps initiated by a snag or a falling bank, the water's inertia carried it to the outer bank, where it flowed faster, while on the inside, the slower moving water deposited silt and sand. The faster moving water was eventually forced to the opposite side, where the process began to repeat itself. Ox-bow lakes occur when the flooding river cuts across the neck of the curve. So many Missouri River banks are now covered with rip-rap, the process can no longer occur. Just in the State of Missouri, the river is now 50 miles shorter than it was a century ago, and the water surface area has been cut in half. A series of exhibits at the **DeSoto Bend Museum** compare the benefits of improved navigation and the reclamation of vast tracts of bottomland for farms with the losses suffered by birds and animals as a result of those projects. The lake you see outside the museum window is one of the few left on the Missouri.

Wildlife isn't the only attraction at DeSoto Bend. On the afternoon of April 1, 1865, the paddle-steamer *Bertrand*, bound for the Montana gold fields, hit a snag and sank in eight feet of water. The passengers were saved, but no attempt at salvage was made, and gradually the river covered the hulk with 25 feet of sand. There she lay until 1967, when a group of salvers, hearing tales of the valuable mercury aboard, started to excavate. Unexpectedly, the treasure proved not to be mercury—though a few containers of that gold-reducing metal were found—but a cargo hold laden with 140 tons of trade goods. It was the historians who reaped the prize, because her varied cargo proved to be a time capsule of mining technology and frontier economy of her day. There were fur hats, bottles of French champagne, canned oysters, bolts of silk, and case after case of Dr. J. Hostetter's Celebrated Stomach Bitters, a 60-proof concoction probably destined for trade with the Blackfeet. Recognizing the importance of such a find, the Department of the Interior built a climate-controlled warehouse for the artifacts and constructed a visitors center where a few of the more important finds could be displayed. The museum provides a fantastic insight into mid-18th-century frontier life.

The best way north from Desoto Bend is to cross the river to the town of Blair, Nebraska, and then turn right on U.S. Highway 73. About 30 miles north you can detour back into Iowa to visit the **Lewis and Clark State Park**. It was near here, on their homeward journey, that the Corps of Discovery got their first word of what had been happening in the United States while they were gone.

Sept. 3rd., 1806 [Clark] at half past 4 P.M. we Spied two boats & Several men, our party pleyed their ores and we soon landed I was met by a Mr. James Air. our first enquirey was after the President of our country and then our friends and the State of the politicks of our country he informed us that Genl. Wilkinson was the governor of the Louisiana two British Ships of the line had fired on an American Ship in the port of New York, 2 Indians had been hung in St. Louis for murder and several others in jale, and that Mr. Burr & Genl Hambleton fought a Duel, the latter was killed &c.&c.

Winnebago and Omaha Indian Reservations

On a map it looks like the Winnebagos and Omahas own a sizable amount of Nebraska real estate. That isn't quite true, thanks to another misguided government policy. The Dawes Act of 1887 provided for the breaking up of reservations into tracts to be assigned to individual Indians, something that was completely foreign to their way of thinking. The original plan was for the government to hold the land in trust, but the Burke Act of 1906 gave fee-simple patent to each allotment. The land could be leased or farmed or sold by the owner, and, of course, it could be taxed by the state. By the time the land was actually issued to the individual, he was often so heavily in debt to the store and the bank that he subsequently lost the land or sold it to pay his debts.

According to the 1980 census, of Nebraska's 10,000 Native Americans, 1275 Omahas live near Macy, 1140 live on the Winnebago Reservation, and 420 live on the Santee Sioux Reservation 100 miles farther upriver. The balance, nearly 70%, live off the reservation. The Omahas are native to the area. The Winnebagos chose to migrate here; they purchased the land with their own money. A highlight of anyone's trip through the country between here and Portland, Oregon, will be to attend one of the tribal powwows. These celebrations, with their singing, dancing, and native dress, preserve a link to the past and affirm tribal traditions for new generations. The Winnebago powwow usually commemorates the 34th Nebraska Volunteers who, under Chief Little Priest, enlisted in the army to help quell an uprising of Sioux, Cheyenne, and Arapaho in 1866.

Most of the reservations we drive through on our westward trek are located on the worst possible land. Since the Native Americans were not farmers, there was no need, so the wisdom went, to give them arable land. That's somewhat true here, but there are some lovely corn fields tucked into the little hollows. What this land lacks in agricultural value, though, is made up by the sheer beauty of the place. The drive through the two reservations is as pretty as you'll find anywhere in the midwest. The road undulates through a cottonwood forest that occasionally opens up to reveal lovely views of the river. You pass the **Chief Blackbird Wayside Rest,** which commemorates one of the most powerful and sinister turn-of-the-century robber barons. Chief Blackbird extracted a toll from all trappers who came by, and got enough arsenic from one trapper so that he could, and did, poison any of his tribe who dared disagree with his policies. He died of smallpox four years before the Corps of Discovery passed this way. Legend has it that he was buried erect, astride his favorite horse. The grave is unmarked. Such was Blackbird's influence that Clark was moved to write:

August 11th, 1804 [Biddle] Blackbird seems to have been a personage of great consideration; for ever since his death he has been supplied with provisions from time to time by the superstitious regard of the Mahas.

Sioux City

Tragedy occurred when the Corps of Discovery reached present-day Sioux City, Iowa.

August 20th, 1804 [Biddle] Here we had the misfortune to lose one of our sergeants, Charles Floyd. He was yesterday seized with a bilious colic, and all our care and attention were ineffectual to relieve him. A little before his death he said to Captain Clark "I am going to leave you;" his strength failed him as he added, "I want you to write me a letter." He was buried on the top of the bluff with the honors due to a brave soldier.

Sergeant Floyd was from Kentucky, where his father had soldiered with William Clark's famous older brother, George Rogers Clark. He had enlisted in the Corps of Discovery as a private, but his fellow soldiers elected him sergeant. His grave is marked by a tall obelisk, clearly visible from the interstate just south of town. Experts believe he died of a ruptured appendix, a malady that could not have been treated in New York City, much less the wilderness, given the state of medical practice in 1804. The river flowing in from the north is named after him.

One of the last surviving riverboats was the *Sergeant Floyd*. Until recently, she made daily tourist excursions out of Sioux City. Her hulk can be seen forlornly rotting away near the interstate, a sad reminder of the glory days of riverboating. In earlier times, this was hardly half way upriver for the specially constructed mountain boats, which could steam all the way to Fort Benton, Montana, 2500 miles from the mouth of the Mississippi. Now, Sioux City is the end of the line; the Corps of Engineers saw no economic justification for the installation of navigation locks on any of the Missouri dams.

Sioux City started life as a brawling river port that quickly grew to a population of 20,000. Upset at its unsavory reputation, one Reverend George Channing, in 1885, undertook a campaign against the saloons, brothels, and gambling dens, and got himself killed for his effort. The meat-packing industry got its start in a curious way. A riverboat ladened with wheat sank; the wet cargo was spoiled, or so everyone thought. One enterprising citizen found that hogs loved the stuff. Since there was no market for live hogs, he started a slaughterhouse, and Sioux City's economy took off from there.

With a population of 80,000, Sioux City is not exactly huge, but it's the largest town we'll visit until we get to Portland, 2000 miles away. Though the city is isolated from the river by the freeway, it nevertheless has a very nice downtown, tastefully refurbished with brick malls, a new hotel, and pedestrian overcrossings one floor above the traffic. The result is that the larger department stores have stayed downtown rather than fleeing to suburban shopping centers. The residential section has a prosperous look and is the location of an historical museum and an art center. **Central High School,** a fortresslike structure that looks like a prison, is unique enough to be included in the U.S. Register of Historic Places.

Sioux City lies at the point where the river swings west again. Though the change is not abrupt, we leave the Corn Belt and enter wheat country. We also leave the navigable part of the river and head into what many people call the Great Lakes region. The lakes, of course, are the reservoirs behind the dams constructed by the Corps of Engineers. Ahead lies the land of the Arikara and Sioux, whom the explorers will meet shortly.

THE GREAT LAKES

Upstream from Sioux City, the Missouri turns more or less westward for about 120 miles. For much of this distance the river forms the boundary between Nebraska and South Dakota. It's along this stretch of river that you leave the Corn Belt behind and enter the great wheat-growing prairie. The change isn't abrupt; it's nothing like you find in North Dakota, where the Missouri essentially divides the state into corn-farming land on one side, wheat and open range land on the other. Nevertheless, the towns you pass from here on are tiny, and they are spaced farther apart. When this was Dakota territory only 607 votes were cast in a congressional election, and even today, on average, only nine people live in a one-square-mile area of South Dakota. Iowa, by contrast, has over 50, and Missouri over 70. As a consequence, you find fewer things to see and do, but at the same time, the adventures of Lewis and Clark become more and more interesting. These men from Kentucky and Virginia who made up the Corps of Discovery were entering a strange land, full of mystery and adventure. Almost immediately they encountered the home of the devil, or so the Yankton Sioux whom they met believed.

August 24th, 1804 [Clark] In a northerly derection [is] an emence Plain a high Hill and by the different nations of Indians in this quarter is Suppose to be the residence of Deavels. that they are in humnan form with remarkable large heads, and about 18 Inches high, that they are very watchfull and are arm'd with Sharp arrows with which they Can Kill at a great distance.

Even though the weather was insufferably hot the explorers couldn't wait to investigate. They found no devils, but they liked the view.

Aug 25th, 1804 [Clark] at 12 oClock we arrived at the hill Capt Lewis much fatigued from heat Several of the men complaining of Great

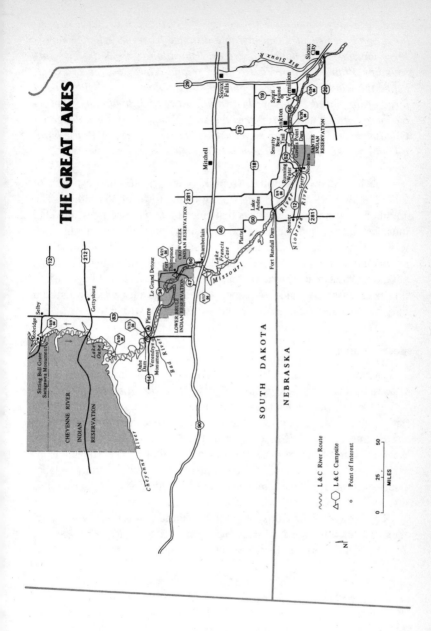

thirst, The only remarkable Charactoristic of this hill is that it is insulated or Seperated in a considerable distance from any other, which is verry unusial in the natural order or disposition of the hills. The Surrounding Plains is open Void of Timber and leavel to a great extent, hence the wind from whatever quarter it may blow, drives with unusial force of the naked Plains we beheld a most butifull landscape: Numerous herds of buffalow were Seen. the Soil of those Plains are delightful Great numbers of Birds are seen Such as black bird, ren, or Prarie burd, a kind of larke about the size of a Partridge.

The hill, known as **Spirit Mound,** is on private land just west of State Highway 19, about 8 miles north of the town of Vermillion. The common wisdom of the day was that since there were no trees, the land must be fallow, and therefore no one could possibly live here, much less farm the land. Where would you get wood to build a house and how could you build a fire? The lack of trees must mean that this is the edge of "The Great American Desert," news of which had begun trickling back east. Like much common wisdom, the idea was wrong; you only have to count the grain elevators to see how wrong.

Vermillion

This pleasant town of 10,000 is the home of the University of South Dakota. Located on the campus is the **W. H. Over Museum,** which has a fine exhibit of South Dakota's wildlife and a section describing the early cultures of the Upper Great Plains. Sharing the same building is the **Shrine to Music Museum,** a treasure trove of 2500 rare antique musical instruments from all corners of the globe. The collection includes pianos, melodeons, bugles, music boxes, lutes, drums, elaborately decorated violins, tambones shaped like dragons, a zither built as a crocodile, and a two-horned trumpet.

The voyage of the Corps of Discovery through this country was anything but uneventful. Patrick Gass was appointed sergeant to replace Sgt. Floyd and 17-year-old George Shannon got lost.

Sept 11th, 1804 [Clark] the Man who left us with the horses 16 days ago George Shannon, joined us nearly Sarved to Death, he had been 12 days without any thing to eate but Grapes &n one Rabit, which he Killed by shooting a piece of hard Stick in place of a ball. This Man Supposeing the boat to be a head pushed on as long as he could, when he became weak and feable deturmined to lay by and waite for a trading boat, which is expected, Keeping one horse for the last resorse, thus a man had like to have Starved to death in a land of Plenty for the want of Bullits.

The Yankton Sioux, unlike their northern cousins the Teton Sioux, were cordial and greeted the party warmly. Clark can't quite decide how to spell *Sioux*. His late August journal entries contain his first reference to a tipi.

Aug 29th, 1804 [Clark] the Scioues Camps are handsom of a Conic form Covered with Buffalow Roabs Painted different colours and all compact & handsomly arranged, Covered all round an open part in the Centre for the fire, each Lodg has a place for Cooking detached, the lodges contain from 10 to 15 persons

August 30th, 1804 [Clark] The Souex is a Stout bold lookin people & well made, the greater part of them make use of Bows & arrows the Warriers are Verry much deckerated with Paint Porcupine quils & feathers, large legins and mockersons, all with buffalow roabs of Different Colours.

The four-lane road west toward Yankton wends through pretty country. You're not in range land yet. You won't see your first sagebrush until you reach North Dakota, but the farms seem to be getting bigger. Many are irrigated with sprinklers mounted atop pipes borne on giant wheels. Most use the pivot system; the whole rig rotates about a center stanchion, where a diesel engine pumps over 50,000 gallons a minute from a pipe bringing water from a nearby dam. Over 125 acres are watered with one rotation of the system. Irrigation produces a larger and more highly valued crop; a field that used to be volunteer grass is now planted to alfalfa, and wheat is replaced by corn stocks yielding 150 bushels to the acre. Land is selling for $3000 an acre hereabouts, so farmers have to go for the big-money crops.

On your trek west it is interesting to see the different methods farmers use to gather and store hay. In these parts, alfalfa is baled by a relatively new type of machine that rolls the hay up into 5-foot-diameter spools that are then carted around by means of a forklift-type attachment on the front of a tractor. Later on you'll see hay piled into the shape of a giant loaf of bread, mounded into a traditional 20-foot-high haystack, baled with a machine that makes 2-by-4-foot cubes, or simply raked into rows and allowed to dry.

Yankton

When Congress created the **Dakota Territory** in 1861 Yankton was made the capital. This territory was larger than Texas, incorporating all of present-day North and South Dakota, Montana, Wyoming, and parts of Idaho and Nebraska. Nowadays Yankton isn't even the capital of South Dakota. When the two Dakotas were granted statehood in 1883—no one knows which was admitted first since President Cleveland insisted the names be hidden when he signed the documents—the seat of government was moved to Pierre. If you fold a map of South Dakota in quarters, first lengthwise, then up and down, Pierre comes out at the junction of the two creases. That's why it's the capital. Though Yankton lost its chance for glory, the 12,000 people who live here now seem to live a warm and prosperous life. The **Dakota Territorial Museum** is housed in the restored Territorial Legislative Council building. You can also visit the **Cramer-Kenyon Heritage House,** a Queen Anne–style home with period furniture.

Just west of town is **Gavins Point Dam,** the first of six huge structures built as part of the Pick-Sloan project, which has converted the upper Missouri valley into 800 miles of reservoir, creating what many call the Great Lakes Country. The Corps of Engineers boasts that all the dams together produce enough electricity to serve a city the size of San Francisco. The reservoir behind the dam is named for our heroes. You can get a good sense of the country by taking a gravel road to a lakeside spot named for the Yankton Sioux Chief Smutty Bear, where the state has built a nice hiking trail. Twenty-five different native plants are identified. Another interesting thing to do here is to ride the last paddle-wheel ferry boat on the Missouri. During the summer she crosses the river from Running Water, South Dakota, to Niobrara, Nebraska. The ferry terminal is just south of the village of Springfield.

The Corps of Discovery began to discover all sorts of new-to-them animals, some exotic, some not.

September 14th, 1804 [Clark] Shields killed a Hare waighing 6¼ pounds (altho pore) his head narrow, its ears large i.e. 6 Inches long & 3 Inches Wide his tail long thick & white.

The hare was of course a jack rabbit; the name has been shortened from the animal with similarly shaped ears, the jackass.

September 13th, 1804 [Clark] I Killed a Buck Goat of this Countrey, about the hight of the Grown Deer, its body Shorter the Horns which is not very hard and forks 2/3 up one prong. the Colour is light gray with black behind its ears down its neck, and its face white: Verry actively made, he is more like the Antilope or Gazella of Africa than any other Species of Goat.

September 17th, 1804 [Lewis] I had this day an opportunity of witnessing the agility and the superior fleetness of this animal which was to me really astonishing. I think I can safely venture the ascertion that the speed of this anamal is equal if not superior to that of the finest blooded courser.

Captain Lewis was right, it is the fleetest animal in America—60 miles per hour for short distances—but it's not a real antelope. Zoologists prefer the name pronghorn but then, they're the ones who also want us to use bison instead of buffalo. "O give me a home, where the bison roam, and the deer and the pronghorn play" doesn't sound quite right, which is perhaps why the scientists' views haven't prevailed. The sighting of the antelope occurred near present-day **Fort Randall Dam,** named for an army post built in 1856. Its only claim to fame, a dubious one, is that Sitting Bull was held prisoner here following his arrest in Canada in 1877. All that remains of the fort is the ruins of a church.

Chamberlain

You get a tremendous sense of the prairie as you drive west along the undulating State Highway 50. All roads here go either north-south, or east-west, usually following township lines, so every now and then you slow down, make a right-angle turn, and head out again. It's a peaceful world. Traffic is light and the only road hazards are combines or plows being driven from one field to another. The explorers were entering a land where there were as many wild animals as Speke found when he was looking for the source of the Nile.

September 17th, 1804 [Lewis] this senery already rich pleasing and beatifull was still farther hightened by immence herds of Buffaloe, deer Elk and Antelopes which we saw in every direction feeding on the hills and plains. I do not think I exagerate when I estimate the number of Buffaloe which could be comprehended at one view to amount to 3,000.

At the little farm community of *Platte,* drive north on State Highway 45 for 25 miles until you come to I 90, where you enter the unique world of the interstate driver. Signs have replaced the buffalo. Captain Lewis's words might be paraphrased for today's driver: "I do not think I exaggerte when I estimate the number of *signs* that could be comprehended at one stretch of interstate to amount to 3000."

- Beautiful Rushmore Cave Black Hills
- A&W Rootbeer Exit 263
- Wall Drug World's Largest Cowboy Orchestra
- Al's Oasis
- KOA Kampground
- Tourism is Our Business—You Be Our Guest Rushmore Bargain Store
- Pioneer Auto Museum Exit 192
- Heart Ranch Family Trails & Rides RV's Welcome
- Black Hills Gold Wall Drug
- Family Land Camper Hookups, Laundry, Showers, LP.
- Wild West Museum 100,000 Items on Display—22 miles.
- 5¢ Coffee Wall Drug
- Gray's Western Ware & Saddlery—Your Country Store
- Flintsone's Campground 265 miles
- We Have It All For You Al's Oasis Three Star Rated
- Old West Pioneer Auto Museum and Antique Town 78 miles
- Oases Inn Free Movie, Golf, Sauna, Whirlpool
- Black Hills Gold Casey's Drug & Jewelry Exit 263
- Free Ice Water Wall Drug

The sign advertising "100,000 items on display" is for a private museum a few miles west of the town of Chamberlain. Among other things, the place is crammed with 100 spurs, 200 bridle bits, branding irons, plows, wagon seats, barbwire, fencing tools, iron toys, pots, rifles, pistols, and washing machines. If you had turned right instead of left on the interstate, you would have been bombarded with signs for the city of **Mitchell**'s famous **Corn Palace,** a municipal auditorium decorated inside and out with varicolored corn cobs. The outside is completely redone each year. A craft's fair is held here when the corn festival is not in session.

In Lewis and Clark's time smoke signals were the chief means of intertribal communication. You see smoke signals along the interstate today, but of a different sort; they come from the eighteen-wheelers, whose drivers find this ideal country to clean out their fuel injectors. They often let fly with an inky cloud that can be seen two states over.

Happily, you turn off the interstate at Chamberlain and head back into the lovely countryside. Chamberlain is a quiet farm community of 2500 people situated on a bluff above the river. The town has a real western feel with false-front wooden buildings. Nearby is the well-tended **St. Joseph Indian School,** with its dormitories housing students who have come off the reservation for their education.

Along about here Lewis became intrigued by the ubiquitous magpie, rarely seen in the eastern part of the country, but familiar to anyone driving these prairie back roads. Magpies feast on the rodents run over by vehicles, savoring each bite until the last second before flying out of range of your car.

September 17th, 1804 [Lewis] the back & a part of the wing feathers are white edged with black, white belly, while from the root of the wings to Center of the back is White, the head nake breast 7 other parts are black the Beeke like a Crow. abt the Size of a large Pigion. a butifull thing.

Lewis seems impressed with what he called a wolf, but what was in fact that lovable, laughable, hateable western nuisance, the coyote.

September 18th, 1804 [Lewis] I Killed a Prarie Wollf, about the Size of a gray fox bushey tail head & ears like a Wolf, Some fur. Burrows in the ground and barks like a Small Dog.

Fort Thompson

One out of every 25 South Dakotans is a Sioux, a fact that is not too surprising when you consider that the word *Dakota* means Sioux. About a thousand live on the **Crow Creek Reservation,** located about 25 miles north of Chamberlain. You don't need billboards to tell you this is the west; your eyes make that abundantly clear. Every little town has its rodeo grounds and often an auction yard, too. HOG SALES EVERY THURSDAY reads a typical sign out front. A flock of sheep graze on one hill, white-face steers on another. A rubber-tired surrey is being pulled by a single trotter. The cowhand (they don't like to be called cowboy) riding in the field doesn't look as you expected; he's wearing one of those plastic brimmed hats with a half-circle hole in the back and a patch on the front saying something like Massey Ferguson. The two Sioux lads in the barnyard are learning the fine art of calf roping. The high-school athletic teams are called the Chiefs. Fort Thompson is the agency headquarters. You can enjoy a sandwich at a restaurant run by the tribe, or buy a six-pack of beer at the liquor store next door. There is money to be made off sportsmen in these parts. The bait shop rents boats and tackle; another store advertises it will freeze your fish, pheasant, duck, and geese.

A few miles out of town, the Big Bend Dam has a visitors center that concentrates on the life and lore of the tribes indigenous to this area. One of the myths they try to disprove is that the natives were always nomads. You learn that it was only after they got horses in the 1750s that the people abandoned their agriculture and began roaming the plains in search of buffalo. **Big Bend Dam** is named after the ox-bow a few miles north of here, which in Lewis and Clark's time was called Le Grand Detour. Captain Clark proved his worth as a cartographer when he painstakingly surveyed the site.

> *September 21st, 1804 [Clark] proceeded on to the Gourge of this Great bend we Sent a man to Measure [step off] the Distance across the gourge, he made it 2,000 yds, The distance arround is 30 miles.*

A copy of Captain Clark's map is displayed in the Big Bend Dam Visitors Center.

Bad River

A few days later Clark's and everyone else's mind was not on geography, but on the feared Teton Sioux. The Corps of Discovery was approaching present-day Pierre, South Dakota, and the most dramatic four days of the entire trip was about to unfold. The explorers had a good idea of what lay ahead. Regis Loisel, a French trapper whom they met a few days out of St. Charles, had told about being humiliated by this tribe, which had extracted a toll that amounted to little more than bribery for permission to pass. If the Missouri was ever to become a trade route to the interior, the Sioux had to be shown that authority and sovereignty lay with the American government; simple river piracy would not be tolerated.

> *September 24th, 1804 [Clark] prepared all things for Action in Case of necessity, The Tribes of the Seauex Called the Teton, is Camped about 2 Miles up on the N.W. Side, and we Shall Call the River after that Nation, Teton.*

Prophetically, Clark's name for the river didn't stick. For reasons that will soon be apparent, it's now called the Bad River.

September 25th, 1804 [Clark] Envited those Cheifs on board to Show them our boat and such Curiossities as was Strange to them, we gave them ¼ a glass of whiskey which they appeared to be verry fond of, Sucked the bottle after it was out & Soon began to be troublesom, one the 2nd Cheif assumeing Drunkness, as a Cloake for his rascally intentions. I went with those Cheifs to Shore with a view of reconsileing those men to us, as Soon as I landed the Perogue three of their young Men Seased the Cable. the Chiefs Soldier Hugged the mast, and the 2nd Chief was verry insolent both in words & jestures declareing I should not go on, Stateing he had not receved presents sufficent from us, his justures were of Such a personal nature I felt My self Compeled to Draw my Sward. at this Motion Capt. Lewis ordered all under arms in the boat, the grand Chief then took hold of the roap & ordered the young Warrers away, I felt My Self warm & Spoke in verry positive terms. We proceeded on about 1 Mile & anchored out off a Willow Island I call this Island bad humered Island as we were in a bad humer.

Perhaps realizing that bluff and bluster would not work, the next day the Sioux tried diplomacy.

September 26th, 1804 [Clark] after Comeing too Capt. Lewis & 5 men went on Shore with the Cheifs, who appeared disposed to make up & be friendly. I went on Shore on landing I was receved on a elegent painted Buffalo Robe & taken to the Village. The great Chief then rose; he then with Great Solemnity took up the pipe of Peace & after pointing it to the heavins the 4 quarters of the Globe & the earth, he made Some disertation, lit it and presented the Stem to us to Smoke,

Apparently, the chiefs felt that flattery didn't work either, for on the following day they again decided on a show of force. They were cautious, though, and well they should have been, for they faced a determined and well-armed adversary. Lewis's decision, made when he was still in Virginia, to strike out with a large party, rather than the small one Jefferson had envisioned, was about to prove propitious.

September 27th, 1804 [Clark] I rose early after a bad nights Sleep found the Cheifs all up, and the bank as useal lined with Spectators. after Brackfast Capt. Lewis & the Cheifs went on Shore, as a verry large part of their nation was comeing in, the Disposition of whome I did not know. The Cheif hollowaed & allarmed the Camp or Town informing them that the Mahars [Omahas] was about attacking them. In about 10 minits the bank was lined with men armed the 1st Cheif at their head, about 200 men appeared and after about ½ hour returned all but about 60 men who continued on the bank all night, the Cheifs Contd. all night with us. This allarm I as well as Capt. Lewis Considered as the Signal of their intentions (which was to Stop our proceeding on our journey and if Possible rob us) we were on our Guard all night, P. C. [Cruzette] our Bowman who cd. Speek Mahar informed us in the night that the Maha Prisoners informed him we were to be Stoped. all prepared on board for any thing which might hapen, we kept a Strong guard all night in the boat, no Sleep.

Again, the Sioux tried a show of force. Captain Lewis, who by this time was justly furious, showed no sign of caving in.

September 28th [Ordway] We then were about to Set off. Some of the chiefs were on bord insisting on our Staying untill the others came. We told them we could not wait any longer. they Sayed we might return back with what we had or remain with them, but we could not go up the Missouri any further. about 200 Indians were then on the bank. Some had fire arms. Some had Spears. Some had a kind of cutlashes, and all the rest had Bows and steel or Iron pointed arrows. Captain Lewis asked the chiefs if they wore going out of the boat. they did not incline to. then Capt Lewis came out and ordered every man to his place ordered the Sail hoisted, then one man went out and untied the chord, which the warrier had in his hand, then 2 or 3 more of their warries caught hold of the chord and tyed it faster than before. Capt Lewis then appeared to be angarry, and told them to Go out of the Boat and the chief then went out and Sayed we are sorry to have you go.

Suddenly, the tribe's resolve collapsed. The mighty Teton Sioux became little more than supplicants.

September 28th, 1804 [Clark] we deturmined to proceed on, the 2nd Chief Demanded a flag & Tobacco which we refused to Give. after much Dificuelty—which had nearly reduced us to necessity to hostilities I threw a Carrot of Tobacco to 1st Chief [A] Warrier came galloping up. we sent by him a talk to the nation stating if they were for peace Stay at home & do as we had Directed them, if they were for war or were De-

turmined to stop us we were ready to defend our Selves. proceeded on about 2 Miles higher up & Came to a verry Small Sand bar in the middle of the river & Stayed all night, I am verry unwell for want of Sleep

Historians disagree on the actual threat the Sioux posed. Bernard DeVoto wrote: "So large a band of Indians—they numbered several hundred and additional ones kept coming in—could easily have massacred the party." The opposite view is held by scientist Paul Russell Cutright, who says: "As we assess the situation, the explorers, 43 strong and armed with the best rifles then obtainable, as well as pistols and swivels, were here pitted against a force of no more than 60 warriors who relied chiefly on bows and arrows." Whether the peril to the Corps of Discovery was real or not seems hardly germane. What is important is that the captains proved their mettle as tough ambassadors from the new government of the Louisiana Territory. It was the Arikaras, not the Sioux, who became the river pirates in later years.

Pierre

Pierre, pronounced Peer, is named for the St. Louis fur trader Pierre Chouteau, Jr., nephew of Lewis's friend. It used to be the third smallest capital in the lower 48 states, but Carson City, Nevada, grew and Pierre didn't, so with less than 10,000 inhabitants, it's now second behind Montpelier, VT. The oft-quoted comment that you can shoot a deer from the capitol steps, though, is simply not true. The **capitol building,** with its ornate dome, is lovely. An interesting mural graces the wall of the governor's outer office. It depicts good and evil, good represented by a white, tomahawk-wielding trapper, evil by a red man, cowering at his feet, about to be clubbed to death. The mural was covered by paint for a while, but a later governor decided it was better to show the bigotry that existed than to hide the offensive painting. Another mural at the top of the main staircase, which depicts smiling Indians trading with fur merchants, is considered more acceptable.

Anyone who thinks Lewis and Clark were the first whites to explore the upper Missouri need only cross the street to the **Robinson Museum** to learn otherwise. A full 60 years earlier a French trapper named Pierre Gaultier de Varennes, Sieur de la Verendrye, explored this area. The following year, 1743, Verendrye's two sons, returning from a visit to the Black Hills, buried a lead plate on the hill west of town. In 1913 some schoolchildren found the treasure, which now occupies a prominent place in the museum. The inscription claims this land for France, a rather presumptuous statement considering that there were a few Arikaras, Mandans, Gros Ventres, Cheyennes, Crows, and Assiniboines who had some other ideas. The Sioux, however, weren't among these tribes; they were still living in Minnesota. A few more years would elapse before the Chippewa, who lived even farther east, and therefore had access to better guns, felt strong enough to drive the Sioux into the Dakotas. The Robinson Museum is interesting for another reason. It houses one of the most complete collections of native artifacts to be found anywhere.

A stone monument marks the site where the Verendrye plaque was found. Cross the river to **Fort Pierre,** turn right before you reach the Bad River, turn right again and follow a gravel road to the top of the bluff. The view overlooking the city of Pierre is lovely. The original Fort Pierre was built near here in 1831, and was, after Fort Union, the most important fur trading post of its day. The artist George Catlin arrived here in 1832 aboard the *Yellow Stone,* as did Prince Maximilian and Karl Bodmer the next year. John J. Audubon spent some time here ten years later. The post was sold to the army in 1855. The commandant, Colonel William S. Harney, signed a treaty with nine Sioux tribes, essentially giving them money if they would let the settlers pass through their lands unharmed. The treaty was widely broken on both sides. The army abandoned the post in 1859.

Fort Pierre observes mountain time, whereas the City of Pierre is in the central time zone. The Missouri River from here to the Montana border forms the boundary, so you will be constantly jumping back and forth between the two time zones. The river is a boundary in more ways than just time, though. It separates the farm land of eastern South Dakota from the range land in the western part of the state. At this latitude at least, Pierre is the western limit of the wheat belt. Those living on the east side of the river look to Minneapolis as their "big city," while those on the west give that honor to Denver.

Lake Oahe

A few miles north of Pierre is Oahe Dam. Nearly two miles across, it is the second largest on the Missouri. Lake Oahe stretches upriver for 231 miles, drowning everything that Lewis and Clark saw between September 28 and October 18. This is the walleye pike capital of the world. The Corps of Engineers boasts that each year more than two million visitors come to camp, swim, water ski, picnic, sail, and fish in this area.

From time to time Lewis and Clark met trappers along the river who told some exciting things about the country that lay ahead.

> *October 1st, 1804 [Clark] Mr. Jean Vallie informs us that he wintered last winter 300 Leagues up the [Cheyenne] River under the Black mountains, The black mountains he Says is verry high, and Some parts of it has Snow on it in the Summer great quantities of Pine Grow on the Mountains. on the Mountains great numbers of goat, and a kind of anamale with large circular horns, this animale is nearly the Size of an Elk.*

Mobridge

U.S. Highway 83 goes northeast from Pierre and then turns due north, following a spur of the Chicago and Northwestern Railroad. The only real settlement in the 100 miles between Pierre and Mobridge is **Selby**, population 900. A sign erected by the town fathers attests to the loneliness of this country. It reads "Wanted—One million visitors." The Corps of Discovery spent five days near Mobridge, where they met the Arikaras, a Caddoan-speaking tribe who were cousins of the Pawnees, but probably had emigrated northward from Santa Fe. Lewis and Clark, who seemed to have mixed feelings about the tribe, referred to them as Rees or Ricaras.

> *October 12th, 1804 [Clark] The Nation of the Rickerries is about 600 men able to bear arms a Great perpotion of them have fusees they appear to be peacefull, their men tall and perpotiend, womin Small and industerous, raise great quantities of Corn Beens Simnins [squash] & Tobacco for the men to Smoke they collect all the wood and do the drugery as Common amongst Savages.*

> *October 9th, 1804 [Clark] three great Chiefs and many others Came to see us to day, much astonished at my black Servent, who did not lose the opportunity of [displaying] his powers of Strength &c. &c. this nation never Saw a black man before.*

> *October 9th, 1804 [Biddle] By way of amusement he told them that he [York] had once been a wild animal, and caught and tamed by his master; and to convince them showed them feats of strength which added to his looks made him more terrible than we wished him to be.*

October 11th, 1804 [Clark] Those people are Durtey, Kind, pore, & extravigent. [They] gave us to eate bread made of Corn & Beens, also Corn & Beans boild. a large Been which is rich & verry nurrishing also Squashes &c. all Tranquillity.

Perhaps the Rees had learned by the "prairie telegraph" that Lewis and Clark were not to be pushed around, because 19 years later emissaries of the Rocky Mountain Fur Co. would find things anything but tranquil. General William Ashley was leading a party upriver to restock the trappers under the direction of his partner, Andrew Henry. He hoped to continue farther up the river to trade with the Blackfeet. They paused here to swap some trinkets for some horses. At first the Rees seemed cordial, but at dawn the next morning they attacked, killing 12 of Ashley's men. One of the survivors, 24-year-old Jedediah Smith, had to swim for his life. An older Jed found himself in a similiar fix on numerous other occasions. For a while the great mountain man seemed to live a charmed life. When the Kelawatset tribe attacked his party of 19 near the Oregon coast, he and three others were the only ones who escaped with their lives. His luck ran out, however, when in 1831 he was ambushed by a band of Comanches on the Cimarron River while en route to Santa Fe.

When news of the massacre of Ashley's men reached Fort Atkinson, the young commandant, Col. Henry Leavenworth, sought revenge. With a party of 250 soldiers and a Howitzer, he ponderously moved upriver in keelboats. Wisely, he recruited over 700 Sioux who were, at the time, itching to do battle with their old enemies, the Rees. When the day for revenge came, the wildly shouting Sioux drove the Rees into their village, but then saw the overcautious Leavenworth fail to press his advantage. He brought up his Howitzer, which on the first shot beheaded the Arikara Chief, Grey Eyes. Still the young colonel hesitated, the Sioux lost interest and drifted away, and the Rees, sensing a chance to escape, sued for peace. Leavenworth demanded reparation of the goods stolen from Ashley's men, which the Rees agreed to return the following day. Dawn found them nowhere in sight. Many of the mountain men, who would later become legends, watched the spectacle—Hugh Glass, Edward (Cut Nose) Rose, Jim Beckwourth, Jim Bridger, "Broken Hand" Fitzpatrick, and several of the Sublette brothers. They were unimpressed. The battle, or lack of it, was to prove crucial, for it sealed off the river route to the American fur trade. Ashley's men and those that followed used the Platte River to reach the fur country. The Missouri was essentially closed until the coming of the steamboat ten years later.

The site of the battle is across the river from the town of Mobridge, 50 feet below the waters of Lake Oahe. Mobridge, so the story goes, got its name because a telegrapher for the Milwaukee Railway celebrated the completion of the bridge across the river by wiring headquarters "Mo Bridged."

This town of 4000, thanks to its excellent fishing and pheasant hunting, is a sportsmen's center, with numerous motels and restaurants catering to the outdoorsman. The old downtown hotel has been converted into "Sportsmen's Condominiums." The Sitting Bull Stampede, an event on the professional rodeo circuit, is held here the first week of July. The town boasts the very nice **Klein Museum,** featuring replicas of an old-time schoolroom and doctor's office, and a collection of Indian artifacts found during archaeological digs conducted just before the filling of the lake. Handmade Indian beadwork and pottery can be bought in the gift shop.

U.S. Highway 12 crosses Lake Oahe and enters the **Standing Rock Indian Reservation.** Turn left on a gravel road that leads to two monuments erected on a bluff overlooking the lake. One, an obelisk, commemorates Sacagewea, the celebrated "Bird Woman" of the Lewis and Clark expedition. The inscription reads in part: "Through her courage, endurance, and unerring instinct, she guided the expedition over seemingly insuperable obstacles." We'll learn, as we head west, that Sacagawea was a true heroine, but she was not a guide. That myth got its start when Eva Emery Dye wrote an historical novel, *The Conquest,* in which she deliberately made Sacagawea bigger than life in order to further the cause of women's suffrage. In South Dakota her name is spelled with a *j,* whereas North Dakota uses a *k.* Sakakawea is closer to the true spelling than Sacajawea, for there was no *j* in the Shoshoni language, but Lewis and Clark scholars insist it should be Sacagawea, and pronounced Sa-COG-ah-we-ah.

The other monument is a sculpture marking the possible burial site of Sitting Bull, who was murdered about 20 miles northeast of here in 1890. For seven years the great chief had been confined to the Standing Rock Reservation, where he saw more and more indignities cast upon his brethren. Many of the tribe, particularly the young men, began to espouse the ritual of the "Ghost Dance," wherein, with much chanting and ceremony, a Messiah was summoned forth to rid the country of the demon whites. The dancers wore a sacred white ghost shirt that would turn away enemy bullets. The longer the incantations went on, the more frightened the white onlookers became. The government Indian agent insisted that Sitting Bull give up the religion and outlaw the practice,

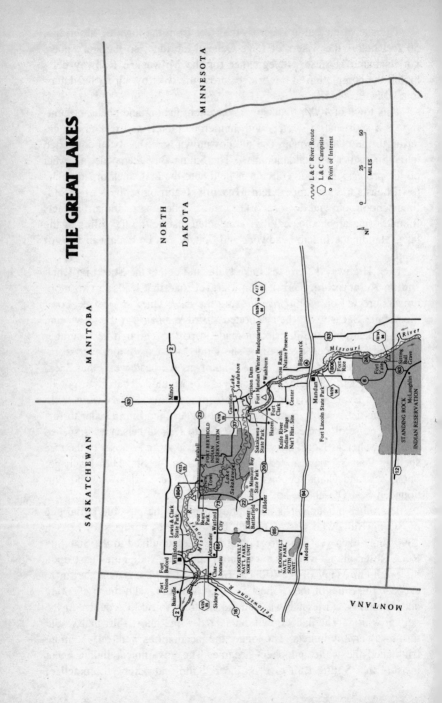

which he refused to do. Forty-three Sioux police, supported by 100 U.S. cavalrymen, were sent to put him under arrest; his followers tried to resist, and the ensuing melee left the great warrior, his son, and six other Sioux dead, killed by members of their own tribe. The rest of the dissidents were herded off to the Pine Ridge Reservation in the southwest corner of South Dakota, where, a few weeks later, most were slaughtered at the battle of Wounded Knee. Three hundred Sioux died in that massacre, two-thirds of them women and children.

Fort Yates

The drive north through the Standing Rock Indian Reservation is through pleasant, undulating hills. Turn off U.S. Highway 12 at McLaughlin and head north on State Highway 63. Along this stretch of road you come to another milestone on your westward trek; you spot your first sagebrush. At the border between North and South Dakota, the road takes a little jog, no doubt because the township lines of one state don't exactly line up with those of another. The land surveyors of the 1890s used state lines as a convenient place to dump all their accumulated errors. A few miles into North Dakota, turn right on the paved road to the town of Fort Yates. The Corps of Discovery passed this spot in mid-October. It was here that Private John Newman faced a court martial for insubordination, the last that Lewis and Clark would have to conduct. The punishment was 75 lashes, dismissal from the Corps, and relegation to common boatman. The Arikaras, who witnessed the white man's form of discipline, were appalled.

October 14th [Clark] The punishment of this day allarmd. the Indian Chief verry much, he cried aloud (or effected to cry) I explained the Cause of the punishment and the necessity of it which he also thought examples were also necessary, & he himself had made them by Death, [but] his nation never whiped even their Children, from their burth.

Fort Yates, with its tree-lined streets, cafes, and grocery stores, is now the principal center for the 8400 Sioux who live on the reservation. The initial burial site of Sitting Bull is marked by a plaque on the left side of the road, just before you enter town. There is much interstate rivalry about whether the great chief's remains are, in fact, here or on the bluff we visited near Mobridge, South Dakota.

Fort Rice

Take State Highway 24 north from Fort Yates and then, when it turns west, continue north on route 1806. There is significance to that number, of course. North Dakota has numbered the roads on the right bank of Lewis and Clark's route 1806, the left bank 1804. Twenty-five miles north of Ft. Yates you come to a partial reconstruction of Fort Rice, where much of the drama that led to Sitting Bull's murder began. By 1865 the Civil War had ended and the army could now turn its attention to the "Indian problem." General Alfred H. Sully established his headquarters at Fort Rice. Apparently, he didn't know about the importance of vegetables in one's diet; 37 of his soldiers died of scurvy the first winter. In 1868, with the assistance of our friend Father De Smet, whom we met at St. Joseph, Missouri, the government induced the Sioux chiefs to come in to sign a treaty, which forever ceded to them all the land between the Platte, Missouri, and Yellowstone rivers. This immense tract included all of western North and South Dakota and a good portion of Montana, Wyoming, and Nebraska. Sitting Bull, despite the pleadings of Father De Smet, refused to sign and instead sent Gall, his lieutenant, to Fort Rice with the message: "Move out the soldiers, and stop the steamboats and we shall have peace." Six years later, in 1874, Lt. Colonel George Armstrong Custer confirmed the discovery of gold in the Black Hills, smack in the middle of the land that was forever Sioux, and suddenly all bets were off. In two years Custer would meet Sitting Bull at a place we will visit later on in this odyssey, the Little Big Horn. Before that fateful day, however, we pick up the trail of Col. Custer at our next stop, Fort Abraham Lincoln.

Fort Abraham Lincoln State Park

Lewis and Clark observed their first earth lodges at a recently abandoned Mandan village in what is now Fort Abraham Lincoln State Park. They also saw, but had not yet learned to fear, the white or grizzly bear.

> *Oct 20th [Clark] Camped on the L.S. above a Bluff containing coal of an inferior quality, After brackfast I walked out on the L. Side to See remains of a village on the Side of a hill. the Chief with Too ne tels me that nation lived in two of villages 1 on each Side of the river and the Troublesom Seaux caused them to move to the place they now live. [Ft. Mandan] The Countrey is fine, the high hills at a Distance with gradual*

Sakakawea: One of the numerous statues of the celebrated "Bird Woman" found along the Lewis and Clark Trail. Photo courtesy of State Historical Society of North Dakota.

assents. Great numbers of Buffalow Elk & Deer, Goats. our hunters killed 10 deer & a Goat to day and wounded a white Bear.

A number of the earth lodges have been restored in the **Slant Indian Village,** so called because the site is on the side of a hill. These are quite likely the remains of what Lewis and Clark saw. You can see for yourself what Karl Bodmer painted and Captain Clark described.

October 12th [Clark] [They] Live in warm houses, large and built in an [octagon] form forming a cone at top which is left open for the smoke to pass, those houses are Generally 30 or 40 foot Diamiter, Covd. with earth on poles willows & grass to prevent the earths passing thro.

The dome house, or earth lodge, illustrates what life was like for these peace-loving farming tribes. You find yourself in a part of North America with little rainfall, about 16 inches a year, which is an important fact, because these dirt-covered houses could not have withstood heavy rains. The houses had a life expectancy of about ten years, which was an incentive for the Hidatsas and Mandans who lived in them to stay at home. These tribes only used the tipi on hunting forays. A fine museum is located near the Slant Village—especially interesting is the slideshow that sensitively illustrates the Hidatsa and Mandan lifestyle.

Directly up the hill from the Slant Village is Fort Abraham Lincoln, which in the 1870s housed six cavalry and three infantry companies. For the common soldier it was a dull life, with nothing to do for months on end. The officers had it better; Mrs. Elizabeth Custer, wife of the colonel, was known for her balls and musicals. She was no doubt apprehensive, however, on May 17, 1876, when, with the band playing "The Girl I Left Behind Me," her husband set out to fight the Sioux. Seven weeks later the steamboat *Far West* sailed downstream with what was left of the Seventh Cavalry, and she and the rest of the world learned what had happened at the Little Big Horn.

Fort Abraham Lincoln occupies a lovely site on a bluff overlooking the river. Several of the blockhouses have been restored, and from their balustrades you have an incredible view of the city of Bismarck, with its prairie-skyscraper, the state capitol, soaring above the undulating plain.

North of Fort Lincoln is the town of **Mandan,** site of the privately run **Great Plains Museum,** which houses a collection of antique firearms and wax statues of Sitting Bull, the Mandan chiefs, and soldiers from Fort Lincoln.

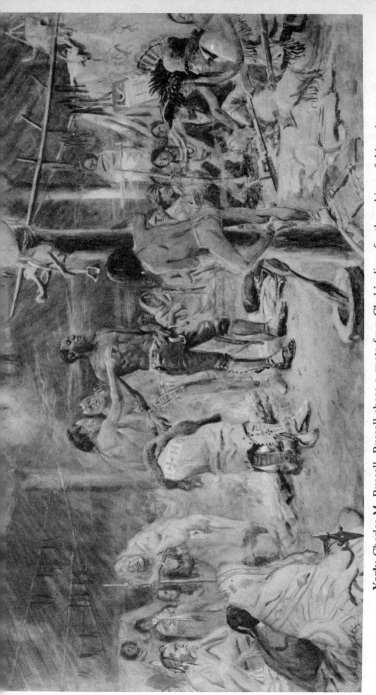

York: Charles M. Russell. Russell chose a quote from Clark's diary for the subject of this painting. "Those Indians wer much astonished at my Servent, they never Saw a black man before, all flocked around him & examind him from top to toe, he Carried on the joke and made himself more turribal than we wished him to doe." Clark and Lewis are seen at the right. Courtesy of Montana Historical Society, Gift of the Artist.

Bismarck

It was money, or rather lack of it, that gave Bismarck its name. The year was 1873, and the Philadelphia banker Jay Cooke was trying to build a railroad. He had convinced—some say bribed—Congress into giving him an enormous land grant as an incentive to push the iron horse from the Great Lakes to the Pacific. The Northern Pacific Railroad was to receive every other section (one square mile) of land for 20 miles on either side of the right-of-way, much of it land owned by the Crow and the Sioux. That was more land than all of New England, with the state of Maryland thrown in, too. The problem was that Cooke couldn't sell the land fast enough to pay construction costs. He induced the one-armed major, John Wesley Powell, discoverer of the Grand Canyon and first Secretary of the Interior, to come here, and Powell subsequently wrote glowingly about the opportunities to be found on the great plains. Jay Cooke, meanwhile, sent agents all over northern Europe trying to promote immigration and settlement. "Plant a nail in the ground and by morning it will grow to be a crow-bar," was the boast.

Even so, few people were buying. In a desperate, last-ditch effort to woo German participation, Cooke promised to name the most important city on the railroad after the Iron Chancellor himself. Otto von Bismarck never raised the money, but the name somehow stuck. Enough settlers did arrive, however, to give this part of North Dakota a decidedly German character. By contrast, 100 miles farther north, in country developed by the Great Northern Railway, the settlers were recruited primarily from Scandinavia.

The rails arrived at the banks of the Missouri in 1873, but they promptly turned to rust, for Cooke's empire had collapsed. He wasn't alone; the Credit Mobelier scandal, involving U. S. Grant's vice-president, Schuyler Colfax, had brought a stop to all railroad construction. Nine more years would elapse before the Northern Pacific crossed the Missouri and made its way on to Seattle. Even then, the owners were still more interested in the land-grab than in running a proper railroad, a fact that would have important land use ramifications for generations to come. The old railroad depot in Bismarck is now an absolutely gorgeous restaurant. Amtrak stopped calling here when Stewart Symington, who had managed to get two routes through his state of Missouri, left the Senate.

The city of Bismarck is an oasis in a sea of wheat fields. Over 45,000 people live here, making it North Dakota's second largest city. Fargo, located on the Minnesota border, is the largest. There is no farm town feel here; you see lots of joggers and bicyclists. The river, which runs free along here, is the summer playground for fishermen, water skiers, and speedboat enthusiasts. All those sandbars that gave Lewis and Clark such fits are put to good use; they are the beaches of the upper Missouri and are crowded with swimmers and sunbathers. The city boasts a lovely park along the river that is home to the charming **Dakota Zoo,** open from May through September. Some of the animals Lewis and Clark first described are on exhibit, including the big-horn sheep, pronghorn, and mountain goat. You also see buffalo, deer, llama, Texas long-horns, and what is billed as the world's largest Kodiak bear. Nearby, a replica of the steamboat *Far West* makes 1½-hour excursions during the summer.

Bismarck now has two shopping centers, one north and one south of town. The result is that the old commercial district is losing its luster. Several old buildings, however, have been restored, including an old warehouse that used to service the steamboat trade.

The capitol is unique in that it looks more like an office building than a typical seat of government. The 19-story structure, finished in a lovely white Indiana limestone, was built at the height of the Great Depression to replace the previous capitol, which was destroyed by fire. The building is open to visitors, and free guided tours are held during the week. The most prominent statue on the capitol grounds is one sculpted in 1910 by Leonard Crunelle of the "Bird Woman," carrying her infant son. (See page 74.)

Showplace of Bismarck, and indeed the entire state, is the brand-new **North Dakota Heritage Center,** located on the capitol grounds. This 127,000-square-foot museum has beautiful, informative displays of the state's history from prehistoric times to the Great Depression, and has extensive exhibits of the local flora, fauna, and geology. The exhibit is as big and as well done as the one beneath the Jefferson Westward Expansion Arch in St. Louis.

Cross Ranch Nature Preserve

The Missouri no longer meanders because the dams prevent flooding, but nevertheless it is nice to find a spot where the river at least runs free and the countryside looks much as it did in 1804. Such a place is Cross Ranch, a Nature Preserve recently purchased by the Nature Conservancy. Although not a formal park, the grounds are open to the public during daylight hours. Self-guided trails skirt 8 miles of riverfront, where you'll perhaps see wild turkey, ring-neck pheasant, antelope, bald eagles, and an occasional bobcat. To reach the ranch, take State Highway 25 northeast from Bismarck. Eighteen miles from the interstate, turn onto Hensler Road and drive north for 5 miles, and then turn east 4½ miles to the ghost town of *Sanger*. These roads are gravel. *Hensler* road continues on to the town of that name, which is near your next stop, Fort Mandan.

Fort Mandan

The maps in existence at the time Lewis and Clark began their trek toward the Pacific primarily illustrated the course of the rivers, for they were really the only landmarks of the day. Eighteenth-century cartographers, like Aaron Arrowsmith, a copy of whose map the explorers took with them, certainly could not show cities or villages, for the plains tribes for the most part never stayed put. There was one exception. All the early explorers and trappers knew of the existence of the Mandan villages. The Mandans were one of the few tribes that stayed in one place. They had a history of being friendly toward the whites, and as a result, their villages became sort of a rendezvous for the French, Canadians, and English who, interestingly, came here to trade for corn, not furs.

It is likely that the captains decided, even before leaving St. Louis, that they would winter here, because they suspected that other trappers would frequent the place, and from them, they felt they had their best chance of learning more about the country that lay ahead. They were right. During the course of the winter they were to encounter at least 15 other white men, who came here the easy way. They canoed across Canada's precambrian shield, an area where the retreating glaciers left thousands of lakes, interconnected by placid rivers. No current to fight here. The men of the Northwest Company and its rival, the Hudson's Bay Company, had only to work their way to Lake Winnipeg and then ascend the Assiniboine and Souris rivers, which brought them within a few dozen miles of the Mandan villages. As we saw in Pierre, South Dakota, La Verendrye led the way, and by the time Lewis and Clark arrived on the scene, the Mandans had seen hundreds of trappers. One wonders why they were so hospitable, considering that the whites had introduced them to smallpox in 1782.

The Corps of Discovery had been on the river 166 days when they decided to make their winter camp. By their reckoning, they had traveled 1600 miles, every one of them against the Missouri's mighty current. It's a bit startling to consider, but the National Park Service estimates that by the time Lewis and Clark arrived here, more people lived along the Missouri River in North Dakota than reside here today. Captain Clark estimated that there were 1250 Mandans, 2500 Hidatsas (also called Minnetaree, Gros Ventres, or Big Bellies), and 200 Wattasoons (also called Ahnahaway, Shoe, and Soulier).

In the early days of November the Corps of Discovery set about making camp. Sergeant Patrick Gass was the principal carpenter.

November 2nd, 1804 [Clark] This Morning at Daylight I went down the river with 4 men to look for a proper place to winter proceeded down the river three miles & found a place well Supld. with wood & returned,

November 3rd, 1804 [Clark] we commence building our Cabins,

November 10th, 1804 [Clark] Continued to build our fort.

November 13th, 1804 [Clark] Ice began to run in the river

November 20th, 1804 [Biddle] We this day moved into our huts which are now completed. The works consist of two rows of huts or sheds, forming an angle where they joined each other; each row containing four rooms, of 14 feet square and 7 feet high,

The fort was finished in the nick of time, for on the 28th the river was choked with ice, and it began to snow. That winter Captain Clark recorded temperatures down to 45° below zero, yet the men were apparently comfortable around their roaring fires. The structure was to be short-lived; it was gone, victim of a prairie fire, by the time the explorers returned in 1806. By the time Prince Maximilian arrived in 1833 even the site was lost to the capricious river's meanderings. In 1973 the McLean County Historical Society built a replica of the fort 3 miles west of the town of Washburn. They chose this site because of its resemblance to what Lewis and Clark saw, though in actuality Fort Mandan was probably located about 7 miles upriver. To view the reconstruction, go north from Washburn on State Highway 200 and turn left on the gravel road. The site is a lovely place for a picnic.

The Corps of Discovery spent more than five months at Fort Mandan. They seemed to have plenty to do. The Hidatsas, or Gros Ventres (pronounced Grow Vons), brought corn, pemmican, jerked meat, dried pumpkins, squashes, and beans. Pemmican, incidentally, a food unique to the North American Indians, was lean buffalo meat or venison cut in thin slices, dried in the sun, pounded fine, mixed with melted fat, and packed in sacks of hide. These gifts of food were repaid, largely because Private Shields was a fine ironmonger. His tomahawks, fashioned out of a damaged stove, became prized possessions among the chiefs. The vegetables were of greater value than the captains probably realized. Prince Maximilian, who wintered just across the river 30 years later, almost died of scurvy. A black cook recognized the problem, fed him some onions, and he promptly recovered. For a while game was plentiful, but as the winter wore on the hunters had to go farther afield to find anything to shoot. It is estimated that it took one buffalo or four deer to feed the party each day.

February 4th, 1805 [Lewis] Capt Clark set out with a hunting party consisting of sixteen of our command and two frenchmen. our stock of meat which we had procured in the Months of November & December is now nearly exhausted. Capt. Clark therefore determined to continue his rout down the river even as far as the River bullet unless he should find a plenty of game nearer. the men transported their baggage on a couple of small wooden Slays drawn by themselves, and took with them 3 pack horses. no buffalo have made their appearance in our neighbourhood for some weeks.

The hunting party was so successful they could not bring all the meat home, but when a second party was sent back to retrieve it, the men learned that it had been stolen by the Sioux.

The men of the Corps of Discovery seemed to find ways to amuse themselves. Peter Cruzette, the son of a French father and a Shawnee mother, played his fiddle. It's rather amazing, but somehow he kept it from getting smashed or warped or stolen throughout the whole 2½-year journey. The Mandan women, like the Arikaras before them, were especially friendly to the soldiers, much to the displeasure of Captain Lewis, who had to deal with the ensuing medical problems. Days were spent repairing gear, making hides into clothes, and on one day, rescuing one of the perogues from the icy clutches of the constantly moaning river.

The captains used the long days to learn as much as they could about the river ahead from the French and British trappers, who had been as far upstream as the confluence with the Yellowstone. From them, they no doubt learned about Alexander McKenzie's epic crossing of the Canadian Rockies in 1793. The captains must have had some uneasiness about the existence of these "foreigners" on American soil, for even then trouble was brewing over where the northern boundary of the United States should lie. That dispute, of course, reached its climax in the War of 1812. The Mandans told them about a large river entering from the north (there turned out to be two rivers, which caused a great deal of confusion later on), and about the great falls they would encounter near the land of the Blackfeet. It was during this time that the captains' journals first mention the French-Canadian, Toussaint Charbonneau, and one of his wives, Sacagawea.

November 4th, 1804 [Clark] a Mr. Chauboine, interpeter for the Gros Ventre nation Came to See us, this man wished to hire as an interpiter,

November 11th, 1804 [Clark] two Squars of the Rock mountains, purchased from the Indians by a frenchmen (Charbonneau) came down.

February 11th, 1805 [Lewis] about five Oclock this evening one of the wives of Charbono was delivered of a fine boy. it is worthy of remark that this was the first child which this woman had boarn, and as is common in such cases her labour was tedious and the pain violent; Mr. Jessome informed me that he had freequently administered a small portion of the rattle of the rattle-snake, which he assured me had never failed to produce the desired effect. having the rattle of a snake by me I gave it to him and he administered two rings of it to the woman broken in small pieces with the fingers and added to a small quantity of water. Whether this medicine was truly the cause or not I shall not undertake to determine, but I was informed that she had not taken it more than ten minutes before she brought forth.

Clark notes that the "squar" was purchased by the Frenchman. There is some dispute over whether she actually was purchased, or simply was won in a gambling game. We do know for sure, however, that she was a Shoshoni (Snake), and that she was about 14 years old when captured by the Gros Ventres near Three Forks, Montana. Eighteen months after that memorable birth, at this exact same spot, Captain Clark wrote in his journal:

August 17th, 1806 [Clark] we also took our leave of T. Chabono, his Snake Indian wife and their child who had accompanied us on our rout to the pacific ocean in the capacity of interpreter and interpretess. T. Chabono wished much to accompany us in the said Capacity if we could have provailed upon the Menetarre Chief to decend the river with us to the U. States, but as none of those Chiefs of whoes language he was Conversent would accompany us, his services were no longer of use to the U. States and he was therefore discharged and paid up. we offered to convey him down to the Illinois if he chose to go, he declined proceeding on at present, observing that he had no acquaintance or prospects of makeing a liveing below, and must continue to live in the way that he had done. I offered to take his little son a butiful promising child who is 19 months old to which they both himself & wife wer willing provided the child had been weened. they observed that in one year the boy would be sufficiently old to leave his mother & he would then take him to me if I would be so friendly as to raise the child for him in such a manner as I though proper, to which I agreed &c.

William Clark was true to his word; he raised little "Pomp," as the child was called. The young lad spent five years in Europe, learned several languages, and later served as an interpreter for the U.S. Army. In 1866, at the age of 61, he left his home in the California Mother Lode camp of Auburn and headed toward the goldfields of Montana. On the way he died of pneumonia and was buried near Danner, Oregon. We have a picture of his father; Bodmer sketched him acting as interpreter for Prince Maximilian. The fate of Sacagawea is subject to much controversy. Most historians believe she died "of a putrid feaver" in South Dakota at the age of 24. Another school, however, maintains she lived to be an old woman, dying on Wyoming's Wind River Reservation in 1888. It's interesting to note that Charbonneau was paid $500.33 for his services; Sacagawea, who was of much greater help, received nothing. One other man left the party here.

August 14th, 1806 [Biddle] we were applied to by one of our men, Colter, who was desirous of joining the two trappers who had accompanied us, The offer was a very advantageous one; and, as he had always perfomed his duty, and his services might be dispensed with, we agreed that he might go, The example of this man shows how easily men may be weaned from the habits of civilized life to the ruder but scarcely less fascinating manners of the woods.

No mountain man ever led a more fascinating life in the woods than John Colter, the discoverer of what later became Yellowstone National Park. We'll run across his path again.

Fort Clark

About 5 miles west of Washburn is the site of Fort Clark, one of John Jacob Astor's American Fur Company outposts. The American art-

The Travelers Meeting with Minatarre Indians: After Karl Bodmer. Maximilian and Bodmer (in top hat) are being introduced to the Minatarre by Charboneau, Sacagawea's erstwhile French Canadian husband. Courtesy of the InterNorth Art Foundation/Joslyn Art Museum.

ist George Catlin spent some time here in 1832 and painted extensively. The following fall and winter, Prince Maximilian and Karl Bodmer were the unexpected guests of the factor, James Kipp, who ordered a crude shelter built to house them. Apparently they had a bit of a time of it, because Maximilian's journal is punctuated with comments about the weather:

[Maximilian] by mid-December the noon thermometer readings were never above freezing . . . At night the cold was so intense, that we could not venture to put our hands from our bodies lest they should be frozen . . . ink, colours, and pencils were perfectly useless 'till they were thawed by the fire.

In spite of the cold, Bodmer created some of his greatest portraits here. The costumes he painted were so colorful, Hollywood moviemakers adopted them for all the Indians they portrayed; even Apache warriors were shown dressed as Mandans.

Six years later, on July 4, 1839, the American Fur Company's steamer *St. Peter's* made a call at Fort Clark, bringing a cargo of pestilence. By July 14, the first Mandan had died of smallpox. By August the factor, Francis Chardon, could no longer keep count. Sixteen hundred Mandans lived here in June; by the end of the year there were fewer than a hundred. The title for Bodmer's collection of paintings—"Views of a Vanishing Frontier"—takes on a poignant meaning.

Knife River Indian Village National Historic Site

Archaeologists have done extensive diggings a few miles north of Fort Clark in an area now administered by the National Park Service. Evidence indicates that Paleo-Indian tribes lived here as early as 10,600 B.C., not long after the retreat of the last glaciers. They probably migrated over a land bridge from Asia. Scientists believe that the Mandans moved here from Minnesota in about A.D. 1000. Rings that outline the former lodges can be seen as depressions in the grass. They delineate two Mandan, one Mahawha, and two Minnetaree villages. It is now believed that these tribes served as middlemen in the northern plains trade. With the Crow, Cheyenne, Arapaho, and Kiowa, they would trade corn they had grown for buffalo robes, fur, and meat. These products would then be retraded for manufactured goods with the Assiniboine and Cree who, in turn, had obtained them from the French. The smallpox epidemics seemed to have come in 20-year cycles, lending credence to the theory that although one generation might finally build up an immunity, its offspring did not.

Another fascinating bit of history surrounds the Mandans. Verendrye had reported that they had white skins. Could it be that they were descendants of the Welsh Prince Madoc who had allegedly discovered America in the year 1170? William Clark was apparently aware of a trip made to this area in 1796–1797 by a Welshman named John Evans, who returned convinced there were no Welsh Indians living in North America. Catlin's and Bodmer's paintings seem to support that theory, for there is not a pale face among them.

This and more can be learned by visiting the Park Service's interpretive center on a lovely grassy plain overlooking the river. A small earth lodge has been constructed nearby. A short gravel road leads to the confluence of the Knife and Missouri rivers, where another village was located. A plaque marks the spot where Charbonneau was living with his two wives when he signed on with the Corps of Discovery, thereby getting his name in the history books.

The Trip Resumes

Spring didn't exactly come early in 1805, but the Corps of Discovery was itching to get going. The French "watermen" were sent home in the keelboat, along with Clark's journals and a collection of animal and floral samples that Lewis had collected.

April 7th, 1805 [Lewis] Having on this day at 4. P.M. completed every arrangement necessary for our departure, we dismissed the barge and crew with orders to return without loss of time to St. Louis. At the same moment that the Barge departed from Fort Mandan, Capt. Clark embarked with our party and proceeded up the River. as I had used no exercise for several weeks, I determined to walk on shore. Our party now consisted of the following Individuals. Sergt.s John Ordway, Nathaniel Prior, & Patric Gass; Privates, William Bratton, John Colter, Reubin and Joseph Fields, John Shields, George Gibson, George Shannon, John Potts, John Collins, Joseph Whitehouse, Richard Windsor, Alexander Willard, Hugh Hall, Silas Goodrich, Robert Frazier, Peter Crouzatte, John Baptiest Lapage, Francis Labiech, Hugh McNeal, William Warner, Thomas P. Howard, Peter Wiser, and John B. Thompson. Interpreters, George Drewyer [Drouilliard] and Tausant Carbono [Charbonneau] also a Black man by the name of York, servant to Capt.

Clark, an Indian Woman wife to Charbono with a young child, and a Mandan man who had promised us to accompany us as far as the Snake Indians.

Clark's journal for that day says essentially the same thing, but he defines the duties of two of his new recruits.

April 7th, 1805 Shabonah and his Indian Squar to act as an Interpreter & interpretress for the snake Indians.

It seems clear that, even then, Captain Clark suspected that Sacagawea would be more valuable to the Corps of Discovery than her husband. Charbonneau, incidentally, spoke French, but not English, so his translations had to be reformed into a language the captains could understand, presumably by Peter Cruzette, Francis Labiche, or Baptiste LaPage, a French Canadian who had been recruited at Fort Mandan to replace the discharged John Newman. The Indian Lewis referred to stayed with the party only two days, so 33 people constituted the Corps of Discovery as it headed into the unknown: two captains, three sergeants, 23 privates, two interpreters, one slave, a squaw, and her infant son. Meriwether Lewis was in high spirits that day.

April 7th, 1805 [Lewis] Our vessels consisted of six small canoes, and two large perogues. This little fleet altho' not quite so rispectable as those of Columbus or Capt. Cook, were still viewed by us with as much pleasure as those deservedly famed adventurers ever beheld theirs; and I dare say with quite as much anxiety for their safety and preservation. we were now about to penetrate a country at least two thousand miles in width, on which the foot of civilized man had never trodden; the good or evil it had in store for us was for experiment yet to determine, and these little vessels contained every article by which we were to expect to subsist or defend ourselves. however, as the state of mind in which we are, generally gives the colouring to events, when the immagination is suffered to wander into futurity, the picture which now presented itself to me was a most pleasing one. entertaining as I do, the most confident hope of succeeding in a voyage which had formed a darling project of mine for the last ten years, I could but esteem this moment of my departure as among the most happy of my life. The party are in excellent health and sperits, zealously attached to the enterprise, and anxious to proceed; not a whisper of murmor or discontent to be heard among them, but all act in unison, and with the most perfict harmony.

Garrison Dam

The harmony of the freely flowing Missouri is lost 20 miles upriver from the Fort Mandan/Knife River area. The huge Garrison Dam backs up the river into one of the world's largest manmade lakes. Nearly 25 million acre feet of water is stored in Lake Sakakawea, almost as much as in Lake Meade behind the Colorado's Hoover Dam. It took Lewis and Clark 2½ weeks to make the 250-mile journey to the mouth of the Yellowstone River. Virtually all of the river they saw is gone today, but fortunately there are several places where you can get a good look at the countryside. One is at **Sakakawea State Park,** near the Garrison Dam. Lewis described the same scene you see today.

April 10th, 1805 [Lewis] The country on both sides of the missouri from the tops of the river hills, is one continued level fertile plain as far as the eye can reach, in which there is not even a solitary tree or shrub to be seen.

Lewis was right about the land being fertile, though for a long time people thought the opposite was true. If there were no trees, it was obvious nothing could grow, and there was no way the land could be settled because there was no lumber for houses, nor cordwood for the stove. General William T. Sherman made a survey of the area after the Civil War and reported "the settlements of Kansas, Dacotah and Iowa have nearly or quite reached the Western limit of land fit for cultivation, Parallel 99° of West Longitude. Then begin the Great Plains 600 miles wide fit only for Nomadic tribes of Indians, Tartars, or Buffaloes . . ." This was the Great American Desert. The settlers who came here by the thousands with the building of the railroad put a lie to that idea. They had their bad years with drought and insect infestation, but they had good years, too, and today, on a cash basis, North Dakota is the number-one grain-producing state in the Union. The state ranks first in premium milling hard red spring wheat, durum wheat, malting barley, flax, and rye.

There is a small visitors center in the **Garrison Power House,** which is decorated with a lovely, rather modern mural of Sacagawea with Pomp slung on her back. Across the dam in the town of *Riverdale,* there is another museum illustrating how the dam was constructed.

New Town

There are two ways to reach the next Lewis and Clark site, New Town. One is to drive north on U.S. Highway 83 and then turn left on

State Highway 37. This is duck-breeding country. The lake you cross is named for John. J. Audubon, who came this way aboard the steamboat *Omega* in 1843. You are passing through the westernmost fringe of the glaciated country that gave Minnesota its 10,000 lakes. A great controversy is currently raging about what to do with the water behind Garrison Dam. Agricultural interests would like to see it pumped east, into the Red River of the North Watershed, where a reliable source of irrigation would make an incredibly rich soil even more productive. Environmentalists are concerned about the possibility of a new species of animal life being introduced into a completely different ecosystem. The Red River of the North drains into Hudson's Bay, not the Gulf of Mexico.

The route west follows State Highway 37 for a while, then continues on as Highway 1804.

The other way to reach New Town is to take State Highway 200 west. After you pass two giant power plants, fueled by locally mined coal, you're out in the country, which becomes more and more wild. This is the fringe of the North Dakota Badlands. At the town of **Killdeer,** where General Sully battled some Sioux, turn north on route 22. The State of North Dakota operates a park about 20 miles north of here. The **Little Missouri Bay State Park** is for those who enjoy a primitive camping experience. Trails lead into the wilderness, where you will find juniper, chokecherry, and creeping cedar on the north-facing slopes and yucca, sage, and cactus on the dry south-facing hillsides. Animals likely to be seen are mule deer, coyote, fox, bobcat, golden eagle, and the prairie rattlesnake.

After Highway 22 crosses the Little Missouri River you enter **Fort Berthold Indian Reservation,** whose headquarters is at New Town, so named because the lake flooded out the site of old Fort Berthold. The fort had its beginnings shortly after the smallpox epidemic of 1837, when the Gros Ventres, Arikaras, and Mandans banded together for mutual protection against the Sioux. Calling themselves the Three Affiliated Tribes, they settled at a place called Like-a-Fishhook Village. Fort Berthold was established in 1860 to trade with the tribes and became an army post when the Indian Agency was established in 1868. Twenty-six hundred Native Americans presently live on the reservation.

New Town is an outdoorsman's center, catering to the fishermen who camp along the shores of Lake Sakakawea. The town's coffee shop is not unlike a thousand others across the country, but here the cashier asks you to sign the guest register. Next door at the Stockman's, the barmistress, who must be in her sixties, won't take any guff from anyone, even a burly ranch hand. At any hint of rowdyism, the place echos with, "Hey cowboy, cut that out!" They pay attention when she hollers.

Just west of town, State Route 23 crosses Lake Sakakawea on the Four Bears Bridge. One of the finest viewpoints in North Dakota is located a few hundred yards off the road, just before you come to the bridge. Turn right and drive up the dirt road to the top of the knoll. This is quite possibly the same bluff where Meriwether Lewis stood when he wrote:

April 22nd, 1805 [Lewis] I asscended to the top of the cutt bluff this morning, from whence I had a most delightfull view of the country, the whole of which except the vally formed by the Missouri is void of timber or underbrush, exposing to the first glance of the spectator immence herds of Buffaloe, Elk, deer, & Antelopes feeding in one common and boundless pasture.

Don't be surprised if the wind is blowing. As one local joke has it, a visitor, leaning into the wind, asks a farmer if this weather is typical. "Nope. It'll be like this for four or five days," the farmer replies, "and then it will blow like hell for a while."

April 24th, 1805 [Lewis] The wind blew so hard during the whole of this day, that we were unable to move. Soar eyes is a common complaint. I believe it origenates from the immence quantities of sand which is driven by the wind this sand so fine and light that they are easily supported by the air, and at a distance exhibiting every appearance of a collumn of thick smoke. so penitrating is this sand that we cannot keep any article free from it; in short we are compelled to eat, drink, and breath it very freely.

If you're lucky, a thunderstorm might be moving in from the northeast. Storms here seem to present a drama quite unlike anything found in other parts of the country, mainly because you can see them coming from such a long way off. The sky starts to boil, the wind can't seem to decide which way to blow, and soon you're dashing to the car to escape the rain. The storms don't last; this country gets only 15 inches of rain a year, much of which comes in the summer.

An interesting controversy surrounds the naming of the **Four Bears Bridge.** It's also known as the "Bridge with Nineteen Names" because there was enough jealousy between rival Hidatsa and Mandan chiefs to prevent them from reaching consensus. Eventually it was named for one of the most famous Mandan chiefs. Mato-Tope (accent over both *o*s),

Mato Tope, a Mandan Chief: After Karl Bodmer. Bodmer painted two portraits of the great chief. On one occasion he earned a "coup" by eating from his enemy's pot before slaying him. Courtesy of the InterNorth Art Foundation/Joslyn Art Museum.

Mato-Tope's Representation of his Combat with a Cheyenne Chief: Mato-Tope. The chief was somewhat an artist himself. The combatants seem to prefer tomahawk and knife rather than the rifles lying on the ground. Courtesy of the InterNorth Art Foundation/Joslyn Art Museum.

or Four Bears, was an artist of some repute. His warlike designs, drawn on tanned buffalo robes, were collected by both George Catlin and Prince Maximilian, and several are on display at the Joslyn Gallery in Omaha. He achieved the admiration of his fellow tribesmen when he crept into an Arikara tipi, calmly ate from the sleeping warrior's pot, and then slew him, thus avenging his brother's murder. Bodmer painted the great chief's portrait twice, one in splendid full dress, the other depicting him as a warrior. Across the bridge, the Three Affiliated Tribes have built an interpretive center that illustrates much of the life and lore of these friendly tribes.

Highway 1806 goes west for a while, but after a bit it turns unpaved, so it is best to jog south and pick up Highway 73 to **Watford City.** This prairie town of 2000 people lies near the entrance to the north unit of the **Theodore Roosevelt National Park.** It's named, of course, for the former president, who had a ranch farther south. These North Dakota Badlands are a scenic wonderland, with wind- and water-eroded rock sculpted into myriad fanciful shapes. General Sully, while campaigning against the Sioux in 1864, came through here and described the country as "hell with the fires out . . . grand, dismal and majestic." The National Park Service operates a visitors center and has built a road that skirts some of the more spectacular formations. The north unit of the park is considerably less crowded than the south unit, which straddles Interstate 94, and has thus become North Dakota's principal tourist attraction.

Fifteen miles beyond Watford is the tiny town of **Alexander,** where you will find the **Lewis and Clark Trail Museum.** It houses a diorama of Ft. Mandan, as well as period artifacts from the days of the pioneers. Turn north at Alexander and recross the river. Not far from here Lewis climbed another hill and commented:

April 14th, 1805 [Lewis] this was the highest point to which any whiteman had ever ascended, except two Frenchmen who having lost their way had straggled a few miles further.

Lewis was referring to LaPage, who had just joined the party. He was mistaken. In 1798 the British explorer David Thompson reached the mouth of the Yellowstone, and presumably some French trappers had preceded him because Thompson recorded that the river was called Rochejhone. At this point the Corps of Discovery was dining like kings.

April 17th, 1805 [Lewis] there were three beaver taken this morning the men prefer the flesh of this anamal, to that of any other which we have, I eat very heartily of the beaver myself, and think it excellent; particularly the tale, and liver

April 21st, 1805 [Lewis] Capt. Clark killed a buffaloe and 4 deer in the course of his walk today; and the party with me killed 3 deer, 2 beaver, and 4 buffaloe calves. the latter we found very delicious. I think it equal to any veal I ever tasted.

Williston

The party was nearing present-day Williston, North Dakota, where you'll have to look long and hard to find veal served in a restaurant. None of that dainty food here; this is steak-and-potatoes country; a dozen different cuts are listed on the menus of restaurants with names like Cattlemen's and the Branding Iron. Williston is about as mid-America as any city you'll find. The population is growing; more than 16,000 live here now thanks to recent oil and gas discoveries. There's a mock-up of an oil-well pump in the local park. A glance through the phone book gives a hint at the social and religious life here. The Yellow Pages lists 27 churches, including two Baptist, two Mormon, and, reflecting the northern European heritage of this country, six of the Lutheran faith, but no synagogue or temple. Seven fraternal organizations are shown: the American Legion, Elks, Knights of Columbus, Masons, Moose, Odd Fellows, and Sons of Norway; and several service clubs: the Jaycees, Kiwanis, Lions, Rotary, and the Business and Professional Women's Association. The farmers who tune in to PBS television here watch a program called "Market to Market." Produced in Des Moines, Iowa, it's the network's western version of "Wall Street Week."

Although Williston is now known as the "Spring Wheat Capital" of the country, it began life as a railroad town. The one-eyed Canadian, James J. Hill, "The Empire Builder," named it after his friend, S. Willis James. He thought there already were enough Jamestowns in the country. Hill's railroad, the Great Northern, was a first-class project from the start. There were no more land grabs to be had, because by 1887 Congress was no longer in such a generous mood. As a result, he had to make money hauling freight and passengers, and for this he needed homesteaders, anyone who would agree to live on and work a portion of the 160 acres the government was giving away free. Attracted perhaps by the $12.50 fare from Minneapolis, homesteaders came in droves, mostly from Scandinavia, attracted by the fertile soil. Yim Hill, as these Norsemen called him, felt that if he was to prosper, they must prosper, too, so he kept freight rates low. Thanks to his insistence on having his tracks laid on gentle grades with wide curves, he got products to market in a hurry. The strategy paid off; Hill's Great Northern eventually absorbed the Northern Pacific and the Chicago, Burlington and Quincy, too. The combined railroads are now called the Burlington Northern.

There will always be controversy about the Homestead Act. Passed by Congress during the Civil War in 1862, its title is "An act to secure Homesteads to actual settlers on the Public Domain." One document described the act thus:

Under these provisions a citizen, 21 years of age, male or female and or the head of a family, could file on up to 160 acres of unappropriated **public** *land. Through the settling, cultivation and improving their claim over a period of five years, with witnesses attesting to this fact, the settlers earned clear patent to their claim. For an investment of five years of their life, the government gave them 160 acres of land. A priceless opportunity for the settlers with out the cash to purchase land and for the Government to ensure the settlement of vast,* **empty** *plains.*

Substitute "Indian" for the two boldface words and the reason for the controversy becomes clear. In these parts, 160 acres wasn't enough land to keep alive on, let alone from which to make a living, but that's another question.

Oil and gas wells aren't the only holes in the ground in these parts today. This is Minuteman Missile country. Strangely, the Air Force doesn't try to disguise the underground silos; you can see them a few hundred yards off the main roads, protected only by a wire fence.

Yellowstone/Missouri Confluence

Williston is where we pick up the so-called Hi Line, U.S. Highway 2, which shoots across the prairie like an arrow, never straying far from the Canadian Border. The Hi Line is almost entirely two lane from Duluth, Minnesota, on Lake Superior, to Seattle on the Pacific, but the traffic is sparse, and you can make as good time as on an interstate. Instead of immediately heading west on the Hi Line, though, detour south along some country roads to one of the most historic spots along the river—

the place where the Yellowstone River, which carries 30% more water, gives up its name to the Missouri. Montanans argue that this is not the true mouth of the Yellowstone, nor is St. Louis, for the Mississippi is smaller than the Missouri. The true mouth of the Yellowstone is therefore 100 miles or so below New Orleans. By Captain Clark's reckoning, this confluence is 1900 miles from the Mississippi. Lewis got here first, though:

April 25th, 1805 [Lewis] I ascended the hills from whence I had a most pleasing view of the country, particularly the wide and fertile vallies formed by the missouri and the yellowstone rivers.

The explorers seemed puzzled over how the river got its name. Lewis sent Joseph Fields 8 miles up the river, but reported back that he could find no yellow stones. Joe Fields would have had to go farther than that; most historians think the river got its name from the color of the rocks near the two giant falls in Yellowstone Park, which are almost 500 miles from here. The Corps of Discovery was happy to have reached this major Western landmark.

April 26th, 1805 [Lewis] in order to add in some measure to the general pleasure which seemed to pervade our little community, we ordered a dram to be issued to each person; this soon produced the fiddle, and they spent the evening with much hilarity, singing & dancing, and seemed as perfectly to forget their past toils.

Fort Buford

More than a year later, on August 3, 1806, Captain Clark, who had descended the Yellowstone, arrived back at this confluence. He commented that the area abounded in beaver and otter, which would make it "a judicious position for the purpose of trade." In the years between 1822 and 1895 the builders of no fewer than nine forts took his advice. The sites of two of them have been made into visitors centers. State Highway 1804 leads first to **Fort Buford,** the last to be built and the last to be abandoned. The grassy field looks much as it did when, in 1866, the army tore down the more historic but by then more dilapidated Fort Union and used the wood to build a new post. It was conceived to be a thorn in the side of the Sioux, and that's exactly what it was. General Winfield Scott Hancock visited the spot in 1868 and noted that it must have been exceedingly offensive to the Sioux, because they had held it under almost constant siege since its founding.

In spite of this harassment, the fort never was seriously threatened, and in later years became a rather civilized place. Several frame buildings still remain, one of which is now a museum operated by the State of North Dakota. The building is crammed full of artifacts depicting military life, and exhibits illustrate some of the things that went on during the fort's nearly 40 years of existence. You learn, among other things, that Chief Joseph was held here for a while in 1877, and Gall, Crow King, and Sitting Bull were prisoners in the 1880s.

By far the most interesting sight to see here, though, is the cemetery, for the recently restored headstones reveal much of what life must have been like for the inhabitants.

Herimiah Burrows	Private	Jul 17, 1868	Shot by accident
Abraham Dick	Private	Apr 11, 1873	Rheumatism
James Cummins	Sergeant	Dec 9, 1862	Gastritis
Daughter of Bloody Knife		May 6, 1876	Disease
Son of Big Hand		Jun 1, 1876	Disease
Son of Sitting Bird		Jun 15, 1876	Disease
Son of Left Hand		Jun 26, 1876	Disease
Mary Lambert	Citizen	Aug 4, 1876	Disease
Owl Headress		Oct 17, 1870	Beat to Death
John Renolds	Private	Sep 9, 1871	Killed by Indians
George Newman	Sergeant	Nov 1, 1876	Thyphoid
Andrew Packer	Citizen	May 6, 1872	Killed by Enemy Scout
William Lee	Private	Dec 3, 1877	Consumption
Rose Stuart	Sergeant	Mar 14, 1878	Dropsy
George Langworth	Citizen	Sep 6, 1874	Murdered
James Fitzgerald	Citizen	Apr 12, 1872	Killed by Road Agents
Thomas McWilliams	Private	Mar 8, 1876	Suicide

The closing of Fort Buford in late 1895 ended an era for the United States Army. Sitting Bull was dead. So were Crazy Horse and Gall. Chief Joseph had long since been subdued, and the battle at Wounded Knee was now but a bitter memory in the minds of the few who survived. The so-called Indian Wars were over. Though the army had hardly won a battle, they had won the war, thanks to better arms and equipment and a seemingly inexhaustible supply of new recruits to throw into the fray.

Fort Union

The first people to take Captain Clark's advice about establishing a trading post here were General William H. Ashley, a 44-year-old Virginian who was lieutenant-governor of Missouri, and his partner, Andrew Henry, who had previously been an associate of Manuel Lisa's. They, along with other St. Louis elite, formed the Rocky Mountain Fur Company. Its first year, 1822, was a huge success, but subsequently the trappers were marauded by nearly every tribe on the upper Missouri. The Rocky Mountain Fur Company was eventually out-muscled by John Jacob Astor's American Fur Company, which built the first substantial fort here. Unlike its neighbor, Fort Buford, Fort Union was built for trade, not war. For nearly 40 years this was an outpost of civilization in a savage land. The roster of its traders and their guests reads like a *Who's Who* of western Americana.

Astor's chief factor, Kenneth McKenzie, built the fort in 1828, and presided with such bourgeois splendor that he earned the title of "King of the Missouri." McKenzie, it is said, imported the best French wines for his table, had countless buffalo tongues in his ice house, wore a shirt of chain mail while dining to impress his Indian chief guests, and had pipers pipe him to dinner like the Highland laird he felt himself to be. He and his successor, James Kipp, played host to George Catlin, Nathaniel Wyeth, and Jedediah Smith in 1832, Karl Bodmer and Prince Maximilian in 1833, Father Pierre Jean De Smet in 1840, J. J. Audubon in 1843, Jim Bridger in 1844, and Isaac I. Stevens, the future governor of Washington territory, in 1856. It took Lewis and Clark three weeks shy of a year to get here from St. Louis. By contrast, the Bodmer/Maximilian journey took 75 days, and 10 years later Audubon was able to do it in 48 days and 7 hours; a record, so he claimed.

The fort was as impregnable as a medieval castle, and quite self-sufficient. The Assiniboine, Crow, Blackfeet, and Sioux chose to make trade, not war. Business was brisk and very profitable. Contrary to government policy, McKenzie built a still at Fort Union, so the furs he bought came cheap. By 1834, though, Astor knew when to quit. He wrote from London: "I very much fear beaver will not sell very soon unless very fine. It appears that they make hats of silk in place of beaver." Astor sold his Western Division to the Chouteau family of St. Louis, who profited on the sale of buffalo robes until they, too, were slaughtered almost to extinction. The Chouteaus in turn sold out to the Hudson's Bay Company in 1865.

Fort Union is now a National Historical Site, the grounds operated by the National Park Service. A small museum shows slide shows and movies depicting the history of the fur trade. In recent years a rendezvous has been staged in early July. Local history buffs come and spend a week, live in tipis, and put on musket-shooting and tomahawk-throwing contests.

Missouri/Yellowstone Confluence

A few hundred yards west of Fort Buford is the present confluence of the two rivers. Picnic tables set on a grassy bluff mark this sylvan scene.

August 4th, 1806 [Biddle] The camp became absolutely uninhabitable in consequence of the multitude of mosquitoes; The face of the Indian child is considerably puffed up and swollen with their bites; the men could procure scarcely any sleep during the night, Captain Clark therefore determined to go on to some spot which should be free from mosquitoes Having written a note to Captain Lewis, to inform him of his intention, and stuck it on a pole at the confluence of the two rivers, he loaded the canoes and proceeded down river.

Captain Lewis, traveling down the Missouri, arrived four days later. Finding a fragment of the note, he made haste to catch up. Four days after that, on the 11th, he was shot by one of his own men. Apparently the one-eyed Peter Cruzette was a better waterman than a hunter, for while he and Lewis were stalking game in a thicket, he mistook his captain for an elk and let fire. The ball hit Lewis in the buttocks and thigh, causing much pain. Cruzette, however, had proved his mettle on the trip to the Pacific, and so he became perhaps the only soldier in the history of the United States Army to shoot his superior officer and escape without even a court-martial.

Meanwhile, Clark was still trying to get away from the mosquitoes. On August 11 he met two trappers from Illinois, the first whites he had seen since leaving Fort Mandan.

August 12th, 1806 [Biddle] they were overjoyed at seeing Captain Lewis' boats heave in sight about noon. But this feeling was changed into alarm on seeing the boats reach the shore without Captain Lewis, who they then learned had been wounded the day before, and was lying in the periogue. After giving his wound all the attention in our power, we remained here some time. The whole party now happily reunited, we left the two skin canoes, and all embarked together, about three o'clock in the boats.

Now that Lewis and Clark are happily reunited, you, the modern traveler, must make a choice. You can either proceed up the Missouri River, following the Corps of Discovery's outward and Lewis's homeward journey, or you can retrace, though in a backward direction, Clark's exploration of the Yellowstone. Both routes offer much to see and do along the way. The next chapter follows the Missouri; the one after that, Yellowstone.

THE MISSOURI BREAKS

It took Lewis and Clark four months to traverse the present state of Montana. During that time they met not one human being, white or red. As you cross the border from North Dakota today, the road turns to gravel, and you begin to wonder if you'll meet anyone either, for the country up ahead looks pretty lonely. You're in Roosevelt County, as big as the state of Delaware, but with a population of only 10,000. The first town you come to is *Bainville,* population 250. "I could put everyone from my high school class up in my own house now," a fellow in his late fifties ruefully commented as he sipped a beer in the town's only bar. Yet tiny farm towns like this seem to have a strong sense of togetherness. Next door to the volunteer firehouse there's a community hall and a lovely park dotted with picnic tables.

Poplar

Your route west rejoins U.S. Highway 2 just north of Bainville. Within a few miles you enter the **Fort Peck Indian Reservation,** traditional home of the Assiniboines, and the enforced home of many of the Yankton Sioux. These two tribes never did get along; forcing them onto the same reservation seems to be one more example of our misguided Indian policy. The Indian Agency is located in the little town of Poplar, where a two-story building houses a **tribal museum.** Fifteen miles farther west, a short side road leads to **Lewis and Clark Memorial Park.** In the steamboat days, woodchoppers made their living here by supplying the boats with fuel. Business must have been good; riverboats used 25 cords of hardwood or 30 cords of cottonwood per 24 hours of steaming. It was along this stretch of river that one of the American Fur Company's boats caught fire and blew up. A deckhand, it seems, tapped a keg of alcohol by candlelight with a gimlet. The fumes, the candle, and the 25 kegs of dynamite stored nearby did the rest.

The Corps of Discovery was living off the fat of the land, always finding more interesting things to write about.

April 30th, 1805 [Clark] I walked on Shore to day our interpreter & his squar followed, in my walk the squar found & brought me a bush something like the currunt, which she said bore a delicious froot and that great quantitis grew on the Rocky Mountains.

May 5th, 1805 [Biddle] The wolves are also very abundant, and are of two species. First, the small wolf or burrowing-dog of the prairies [coyote] the second species is lower, shorter in the legs. They do not burrow, nor do they bark, but howl; they frequent the woods and plains, and skulk along the skirts of the buffalo herds, in order to attack the weary or wounded.

Wolf Point

The party was nearing Wolf Point where, today, you will find neither coyote nor wolf. An historical marker near town tells why. "A party of trappers poisoned several hundred wolves one winter, hauled the frozen carcasses in and stacked them until spring for skinning. It taught those varmits a lesson. No wolf has darkened the door of a house in this town since." Wolf Point, with its block-and-a-half-long main street, looks substantially more prosperous than Poplar.

About 25 miles west of Wolf Point you cross the Milk River, which the Corps of Discovery encountered on May 9th.

May 8th, 1805 [Lewis] We nooned it just above the entrance of a large river which disimbogues on the [starboard] side; I took the advantage of this leasure moment and examined the river about 3 miles; the water of this river possesses a peculiar whiteness, being about the colour of a cup of tea with the admixture of a tablespoon of milk. from the colour of it's water we called it Milk River. we think it possible that this may be the river called by the Minitares the river which scoalds at all others.

The Milk was the first of two rivers coming in from the north that the explorers would encounter. The Gros Ventres had only mentioned one, located near the great falls. This led them to believe that they were much closer to the Rocky Mountains than they actually were, and when they finally came to the second river, the Marias, its unexpected ap-

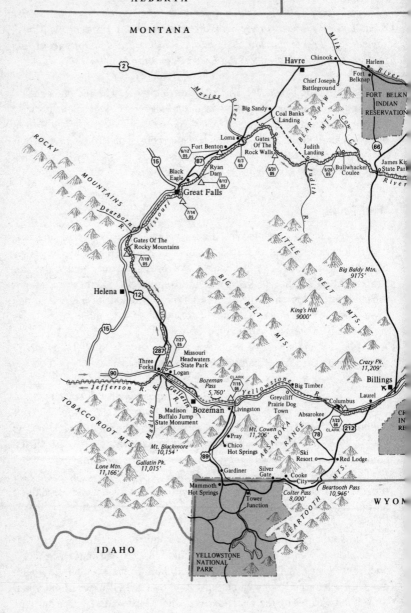

MISSOURI BREAKS

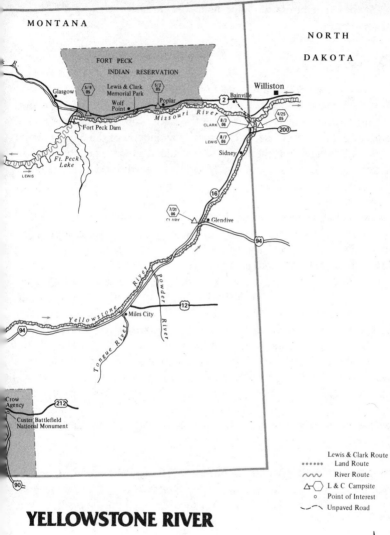

YELLOWSTONE RIVER

pearance put them in such a quandary they lost nine days trying to figure things out. They never really got reoriented until they found the Shoshonis four months later. Incidentally, if Lewis's "leisure" was to walk six miles during the noon hour, one wonders what he did for serious exercise.

Fort Peck Dam

Almost four miles long and two thirds of a mile thick at its base, Fort Peck Dam is huge by anyone's reckoning. Built by the WPA during the Great Depression, it became the cornerstone of the Corps of Engineer's Pick-Sloan Project. The view from the top of the dam is one of a vast wilderness. For the first time on your westward trek, you're now above 2000 feet, and the reason Montana adopted the slogan "Big Sky Country" becomes patently obvious. Whoever first called it "a place where the sky comes down the same distance all around" must have been standing nearby. During the spring the land can become incredibly green—a gentle breeze creates "a sea of grass." Toward summer the green turns tawny, and swift-moving thunderclouds can give even fallow land a sense of motion. Your eyes make you feel as if you are at sea, such is the lack of landmarks and the distance to the horizon, but your other senses tell you otherwise. There is no stickiness to the air; it's dry, not damp, and the smell is not of salt, but is the sweet aroma of a million sagebrush bushes carried by the breeze.

As you look across the windswept waters of Fort Peck Lake, it's not difficult to imagine one incident that occurred about 30 miles upstream from here. Charbonneau was at the helm of the white pirogue when a sudden squall caused the canoe to broach. Eventually she righted herself, but not before water filled to within an inch of her gunwales. Both captains were ashore at the time, watching helplessly as the panic-stricken French-Canadian did nothing.

May 14, 1805 [Lewis] Charbono still crying to his god for mercy, had not yet recollected the rudder, nor could the repeated orders of the Bowsman, Cruzat, bring him to his recollection untill he threatened to shoot him instantly if he did not take hold of the rudder and do his duty.

Somehow they got the perogue to shore, where they promptly set up camp and started drying things out. Later, Lewis, reflecting on the incident, gave credit where credit was due.

> *May 16th [Lewis] the loss we sustained was not so great as we had first apprehended; our medicine sustained the greatest injury, several articles of which were intirely spoiled. the Indian woman to whom I ascribe equal fortitude and resolution, with any person onboard at the time of the accedent, cought and preserved most of the light articles which were washed overboard.*

Sacagawea had once again proved her mettle.

Fort Peck Dam backs up water for 135 miles. The lake flooded the site of the old Fort Peck, a trading post built in 1867, and also swamped one of America's greatest archaeological sites. Between 1907 and 1914, the noted scientist Dr. Barnum Brown found some of the most outstanding fossils ever discovered, most of which were assembled at the American Museum of Natural History in New York City. The highlight was the discovery of the 18-foot-high, 47-foot-long skeleton of the Tyranosaurus Rex, king of the dinosaurs. Some of Dr. Brown's finds are displayed at the powerhouse located at the east end of the dam.

You have to leave the trail of Lewis and Clark here unless you have a boat. A boat provides access to some nice walleye, northern pike, trout, crappie, catfish, and perch fishing, but the things Lewis and Clark saw between here and James Kipp State Park are under water. It's lonely country up ahead. With the single exception of U.S. Highway 191, which darts across the river at Kipp Park, there is not even a paved road that comes anywhere near the Missouri between here and Fort Benton, 265 miles away.

This stretch of river, the Missouri Breaks, is so named because of the landform that has resulted from the downward cutting action of the river. Lewis and Clark, and after them, Maximilian and Bodmer, had some fantastic experiences traversing this great American wilderness, particularly that part of the river, upstream from Kipp Park, which is not under water. You too can have some fantastic experiences, because most of the river between James Kipp Park and Fort Benton has been declared "Wild and Scenic." Motorboats, though allowed, cannot be run fast enough to make a wake, precluding upstream travel. As a result, if you're willing to spend a few days out in this wilderness, you'll not only have a peaceful experience, you'll also have the pleasure of seeing about the only countryside between St. Louis and the Rocky Mountains that looks exactly as it did in 1805. Since this excursion requires downstream travel, we won't pick up the explorers trail again until Fort Benton.

Glasgow

When James J. Hill built his railroad in 1887 he took one look at the Missouri Breaks and concluded that following the nearly parallel Milk River would be a whole lot easier. The highway engineers who built the Hi Line did the same. Driving along Route 2, the first place you come to after Fort Peck Dam is the delightful village of Glasgow. The people of Valley County have put together a lovely **Pioneer Museum,** which is open to the public between Memorial Day and Labor Day. Exhibits and dioramas illustrate the history and culture of the area from the days of Lewis and Clark to the coming of the sod-busters in the 1920s.

A hundred miles west of Glasgow, you enter the **Fort Belknap Indian Reservation,** where the Gros Ventres and Assiniboines live harmoniously together.

Chinook/Chief Joseph Battleground

The little farming town of Chinook is named for the Indian word meaning "warm, thawing winds of late winter." It, too, boasts a museum describing life in the old days. Exhibits include furnishings from a prairie schoolhouse, a doctor's and dentist's office, and a homesteader's shack. The museum also includes a collection of ancient sea-bed fossils found nearby.

Twelve miles south on a narrow country road is the place where the tragic flight of the Nez Perce (pronounced Purse) came to its sad but heroic climax. It was October 1877. For the last 112 days Chief Joseph and his decimated band of women, children, and warriors had marched 1300 miles, time and again outfighting and outmaneuvering the United States Army. They hoped to seek asylum in Canada, where Sitting Bull had recently fled. The starving and exhausted party had paused to hunt for game at the base of the Bear's Paw Mountains, only 40 miles from their goal, when Colonel Miles and the lumbering army of General Howard caught up with them. Cunning and daring generalship would no longer work—the Nez Perce were outnumbered 100 to 1. Miles's terms were simple: "If you will come out and give up your arms, I will spare your lives and send you to your reservation." Chief Joseph's reply was what many believe to be the finest surrender speech ever given:

Tell General Howard I know his heart. What he told me before I have in my heart. I am tired of fighting. Our chiefs are killed. Looking Glass is dead. Toohoolhoolzote is dead. The old men are all dead. It is the young men who say yes or no. He who led on the young men is dead. It is cold and we have no blankets. The little children are freezing to death. My people, some of them, have run away to the hills, and have no blankets, no food; no one knows where they are—perhaps freezing to death. I want to have time to look for my children and see how many of them I can find. Maybe I shall find them among the dead. Hear me, my chiefs! I am tired; my heart is sick and sad. From where the sun now stands I will fight no more forever.

True to form, the army, instead of conducting the Nez Perce to their Idaho homeland as promised, shipped them like cattle to Fort Leavenworth, Kansas. We will follow the flight of Chief Joseph and his heroic band, albeit in a backward direction. We'll see them in action at Big Hole, Lolo Pass, White Bird, and Fort Lapwai; but nowhere does their tragedy have more poignancy than here, 12 miles by country road from civilization at this now peaceful spot at the foot of the lovely Bear's Paw Mountains.

Havre

Havre, pronounced HAVE-er, is a railroad town, the principal division point on the Great Northern between Spokane and Minot, North Dakota. A stroll through the railroad yard gives an idea of what is happening to that industry. Much of the freight you see these days is containers and trailers, mounted on flat cars. The names on the sides give a hint of the cargo: Nippon Lines, Sea-Land, Hapag Lloyd, American President Lines. This railroad, thanks to its low pass over the Continental Divide, is now a land bridge, ferrying seaborn freight between ships calling at Seattle and Portland on the Pacific and at Duluth, Minnesota, on the Great Lakes. Havre is a stop on Amtrak's transcontinental train, the Empire Builder.

Havre is the home of Northern Montana College, which, along with Eastern, at Billings, and Western, at Dillon, form the second tier of Montana's higher education system. The army built **Fort Assiniboine** here in 1879 for the express purpose of preventing the local tribes from fleeing from their "homeland" to Canada. It was a bit like closing the barn door after the horse was gone. More than 10,000 people live here now, and it's the trading center for north-central Montana. Don't be fooled by the bib overalls, cowboy boots, and unassuming look of many of the people you see on the streets and in the restaurants. There is a lot of money in these parts, and the fellow you see wearing his hat at dinner could quite well own a spread as big as a back-east county. A comment you might hear, though, is: "I got out of the army with seven hundred and fifty dollars. Now I'm seven hundred and fifty thousand in debt." One local joke goes: "My neighbor farms ten thousand acres and I farm all around him. I have ten acres up north, thirty acres here and twenty-two over to the west!"

For present-day Lewis and Clark trekkers, Havre is another place where you have to make a decision on how to proceed. The Corps of Discovery was, of course, following the Missouri, which at this point is 50 miles to the south. To retrace their voyage you turn southwest on U.S. Highway 87. On their return trip, however, Meriwether Lewis and three of his best men explored the country near present-day **Cut Bank**, Montana, which is 125 miles due west of Havre on U.S. Highway 2. The country around Cut Bank is spectacular, the area dominated by the craggy peaks of **Glacier National Park.** It's worth seeing, too, so it's best to go to both places. We will, therefore, follow Lewis's return route in the chapter "Blackfeet Country."

Loma

Recall that we left the Corps of Discovery at Fort Peck dam. You now rejoin their trail at the tiny hamlet of Loma, where the Marias (pronounced Mar-EYE-as) River joins the Missouri.

June 8th, 1805 [Lewis] I determined to give it a name and in honour of Miss Maria Wood called it Marias River. it is true that the hue of the waters of this turbulent and troubled stream but illy comport with the pure celestial virtues and amiable qualifications of that lovely fair one; but on the other hand it is a noble river;

Maria Wood was Lewis's cousin. The explorers hadn't expected to see another river coming in from the north and they were confused.

June 3rd, 1805 [Lewis] This morning early we passed over and formed a camp on the point formed by the junction of the two large rivers. here in the course of the day I continued my observations. An interesting question was now to be determined; which of these rivers was the Missouri. to mistake the stream at this period of the season, two months of the traveling season having now elapsed, and to ascend such stream to the rocky Mountain or perhaps much further before we could inform ourselves whether it did approach the Columbia or not, and then be obliged to return and take the other stream would not only loose us the whole of this season but would probably so dishearten the party that it might defeat the expedition altogether.

A quarter of a mile or so after you cross the Marias River, you come to a little gravel road that goes down to the Loma ferry. Turn left, drive up the hill, park, and then climb under the split-rail fence. Being careful about rattlesnakes, walk to the top of a bluff, where you'll find the best view of the confluence of the two rivers. Your immediate reaction will be, "What was their problem?" Thanks to some upstream irrigation projects, the present Marias River has only a tenth the water the Missouri does. But the explorers arrived at the height of the spring runoff and the Marias was muddy, whereas the Missouri was sparkling clear. For nine days the captains reconnoitered the country, Lewis exploring the Marias, Clark the Missouri. They both came to the same conclusion— the clear-watered southern stream was the true Missouri. Not one other member of the party agreed.

It's perhaps fun to second-guess the explorers at this juncture. They were correct in that the Marias was not the true Missouri, but had they followed this river, they would soon have come to Marias Pass, which was the one chosen for the route of the Great Northern Railway. It's the lowest transcontinental railway pass in America. Even though they would have had a time of it descending the V-shaped canyon on the other side, they probably would have saved at least two months in their westward journey.

This strategic spot had its day in the sun. It was the first permanent American foothold in the home of the dreaded Blackfeet Nation. The short-lived Fort Piegan, named after one of the Blackfoot tribes, was built in 1831 by James Kipp. Although he took 2400 beaver plews in the first ten days of trading, his employees refused to spend the winter in this hostile land. They returned to Fort Union, whereupon the Blackfeet set the buildings afire. The following year proved more successful. Kipp constructed Fort McKenzie, named after his boss, the so-called "King of the Missouri." Where Lisa, Ashley, and Henry had failed, McKenzie and Kipp were able to establish trade with the greatly feared Blackfeet, largely because they employed an old Hudson's Bay Company trapper who could speak their language and who was able to explain the advantages of trade. The result was that before the next winter was out, James Kipp had taken some 4000 pelts from the area. For a while Fort McKenzie was the American Fur Company's most profitable

Fort McKenzie, August 28th, 1833: After Karl Bodmer. Bodmer actually witnessed this battle between the neighboring Blackfeet and a band of marauding Assiniboines and Cree. No trace of the fort exists today. Courtesy of the InterNorth Art Foundation/Joslyn Art Museum.

post. It was finally abandoned in 1843 when Fort Benton, 12 miles upstream, took its place.

Bodmer and Maximillian spent six weeks at Fort McKenzie, the farthest outpost of civilization in the Louisiana Territory at the time. Shortly after dawn on the morning of August 28, 1833, Bodmer had the unique opportunity to witness a battle between the vaunted Blackfeet and an army of 600 Assiniboines and Cree. His painting of the melee gives us perhaps the most vivid record ever of the war dress and fighting customs of the plains tribes. Warriors are shown wielding guns, tomahawks, spears, and bows and arrows, and one brave is holding up the bloody scalp of a vanquished foe lying at his feet. During his stay Bodmer also painted memorable likenesses of both Piegan and Blood Blackfoot chiefs.

The tiny village of Loma is all that is left of this once famous spot. The peace between white and native that Bodmer saw didn't last. Many of the Blackfeet were wiped out by the smallpox epidemic of 1838, and hard feelings lasted for a long time afterward. Some Quaker settlers planned to make a new life here in 1865. Their town was to be called Ophir. A work party of ten was summarily massacred by the Blackfeet.

There is one other interesting thing to see here. The State of Montana pays for the operation of about a dozen ferries that cross the river in places that are either too isolated or where the traffic is too sparse to justify the building of a bridge. One of these relics of a bygone age is just a mile off the highway at Loma. The bargelike boat, hardly big enough for a reasonably sized hay truck, is pulled across the river by a winch, cranking in an overhead cable.

Fort Benton

Fort Benton, for a while, occupied an even bigger place in the sun. Lying at the head of navigation for the seemingly endless Missouri, it was the jumping-off point for the gold fields of western Montana, the place where steamboat met ox train. The teamsters and muleskinners who drove the wagons met the deckhands and boilermasters of the boats, and when they got together all hell broke loose. One old history book describes Fort Benton as "a polyglot town of French, Indians, Blacks, wolfers, bullwhackers, whisky runners and a few solid citzens." The town began life as the American Fur Company's replacement for Fort McKenzie. Built in 1846 and originally called Fort Lewis, the company felt it prudent to rename it Fort Benton, honoring their best friend in Congress, Missouri's Thomas Hart Benton. When Montana became a state this county was named after the owners of the American Fur Company; you're in Chouteau county. The fort was built of adobe—unusual for the upper Missouri—and fragments of it still survive. The first steamboat arrived in 1859. The *Chippewa,* carrying 160 tons of freight, steamed back downstream with a cargo of 1800 buffalo robes.

It was gold, though, not buffalo hides and furs, that put Fort Benton on the map. In 1862 some drifters found color in Grasshopper Creek in southwestern Montana, and the rush was on. One boat in 1864 brought a camel train, and another year saw 8000 tons of barrels, boxes, tools, bales, and equipment and 10,000 passengers arriving at Fort Benton's docks. A steamer departed for St. Louis carrying 2½ tons of gold, worth $1¾ million. Twenty-five hundred men, driving 3000 mule teams and 20,000 oxen, freighted supplies to the mines on a road built by one Lt. John Mullan. We'll follow portions of the Mullan Road in later chapters. By 1866 31 steamers had arrived, seven of them at one time. Ten years later the number grew to 60 and there must have been a veritable traffic jam. Commerce was helped by the establishment of the Whoop-up Trail to Canada, where Fort Benton's merchants could buy whiskey to sell to the Indians, an illegal practice in the United States at the time. Then, in 1887, the townsfolk welcomed the arrival of the first Iron Horse, and their doom was sealed. Fort Benton lost its reason for being. Only four boats arrived in 1889, and the last load of river-borne freight sailed downstream in 1890. Some government snag boats came upriver in the early 1900s. Then it was all over.

Although Fort Benton lost its boisterousness, it managed to keep its charm. Seventeen hundred people, living in neat homes on quiet, tree-shaded streets, call this place home. About that many come from outlying areas to shop or to swap stories over breakfast in the morning, or a beer later on. This is wheat country—the county has more than 850 farms averaging nearly 300 acres each. Seventy thousand head of whiteface cattle and a few thousand sheep roam the desolate range land.

Fort Benton's main street, which parallels the river, is about as picturesque as any you'll see in Montana. A number of 1880s buildings house the town's merchants. The pride of the city is the old **Grand Union Hotel,** designated as a National Historical Landmark. The building is undergoing renovation and expects to reopen as the luxury hotel it used to be. A 4-block-long park lines the riverbank. Numerous historical signs describe life in the glory days, and the keel boat used in the filming of the movie *The Big Sky* is displayed nearby. A fine statue by Bob Schriver, a sculptor from Browning, Montana, depicts Lewis and Clark with Sacagawea and her child. It graces the plaza across from the historical museum. The museum, one of the best in the state, has an especially interesting exhibit of the history of the wagon trains, the only museum on the Lewis and Clark trail that does so.

Gates of the Rock Walls

Before leaving Fort Benton, pause long enough to engage the services of an outfitter, and then, instead of heading upriver, go the other way. Congress has declared the 150 miles of river between here and James Kipp State Park to be "Wild and Scenic," meaning no development is permitted. The Bureau of Land Management, a section of the Department of the Interior, administers the area, enforces the rules as to what types of activities are allowed, and maintains a number of campsites along the river. The emphasis is on providing people with a wilderness experience. To see this country, you can either paddle downstream in a canoe or ride a larger boat operated by one of several people who make their living taking people into this spectacular country. One of the best is Bob Singer's Missouri River Outfitters, which schedules numerous 2- to 5-day trips from mid-May through Labor Day. What makes Singer's trips so interesting is that he not only knows the country, but is a Lewis and Clark scholar as well. He brings along the journals and copies of Bodmer's paintings, so you can compare what they saw and did with what you see and do today.

There are a number of places to put in and take out. The trip from Fort Benton to Kipp State Park takes about four days, though it can be done faster. The prettiest section is just below **Coal Banks Landing,** a put-in point at the end of a 5-mile gravel road. There's not much here now, only an isolated farmhouse and the house trailer where the BLM ranger lives. Coal Banks used to be the boat landing for the town of Havre. The coal, incidentally, is lignite, of such poor quality that even the fuel-hungry riverboats couldn't make steam by using the stuff.

As you float downstream you almost immediately enter an enchanted world, a section of the river known as the **White Cliffs.** Meriwether Lewis described the country as well as anyone has since.

May 31st, 1805 [Lewis] The hills and river Clifts which we passed today exhibit a most romantic appearence. The water in the course of time has trickled down the soft sand clifts and woarn it into a thousand grotesque figures, which with the help of a little immagination and an oblique view are made to represent eligant ranges of lofty freestone buildings, having their parapets well stocked with statuary; collumns of various sculpture both grooved and plain, in other places with the help of less immagination we see the remains or ruins of eligant buildings; some collumns standing and almost entire with their pedestals and capitals; As we passed on it seemed as if those seens of visionary inchantment would never have an end; for here it is too that nature presents to the view of the traveler vast ranges of walls of tolerable workmanship

Prince Maximilian, observing the same stretch of river, punctuated his diary with phrases like "a castle or barracks room built upon a summit"; "It was like an old Gothic Chapel with a chimney and pines . . . growing around the wall"; "in a garden laid out in the old French style, where urns, obelisks, statues as well as hedges and trees clipped into various shapes, surround the astonished spectator." Today, the countryside abounds in turtles, beaver, pheasant, white-tailed and mule deer, and sharp-tailed grouse. Antelope roam the prairies, while in the sky above reel pelicans, osprey, and great blue herons. There are some forest trees on the sandstone hillsides, mainly Douglas fir, pine, and juniper, all of which is second growth. The riverbank is dotted with oasislike spots where the ubiquitous cottonwood provides a welcome mantle of shade.

Drifting by, you get constantly changing views of landmarks with names like Fortress Rock, Burned Butte, Grand Natural Wall, Hole in the Wall, and the spectacular cliff Bodmer painted known as the Citadel Rock. You pass milestones familiar to the steamboat captains—Pilot Rock, LaBarge Rock, and Steamboat Rock, the latter of which the boat captains calculated to be exactly 2215 miles from the Gulf of Mexico. Deadman Rapids got its name because the boatmen often had to sink an iron bar (a deadman) in the river, tie a cable to it, and winch the boat upstream. Bob Singer says many a steamboat photo was taken here, for passengers had time to jump ashore and set up their cameras.

Captain Lewis was especially interested in a species of sheep that the Mandans had told him about.

May 25th, 1805 [Clark] I walked on shore and killed a female Ibi or big horn animal The Horns are large at their base, and occupy the crown of the head almost entirely, they are compressed, bent backwards and lunated; the surface into wavey rings which incircleing the horn continue to suceed each other from the base to the extremity. this horn is used by the natives in constructing their bows; I have no doubt of it's elegance and usefullness in hair combs, and might probably answer as maney valuable purposces to civilized man, as it does to the native indians, who form their water cups, spoons and platters of it.

View of the Stone Walls: After Karl Bodmer. Travelers willing to take the time to canoe down the "wild and scenic" portion of the upper Missouri River will be rewarded with views of the exact same cliffs Bodmer painted in 1833. The likelihood of spotting a Bighorn Sheep, however, is small. Courtesy of the InterNorth Art Foundation/Joslyn Art Museum.

Bighorn still roam this country, but your best chance of seeing them is at a place we'll visit in "Blackfeet Country," the National Bison Range, north of Missoula. Just below the White Cliffs you pass beneath a bluff known as the Buffalo Jump. In the late-18th century, the plains Indians, lacking effective weapons, had to resort to subterfuge on their buffalo hunts. One young warrior would be dressed in buffalo skins and sent out as a decoy. He would cleverly place himself at the top of a cliff while the rest of the party would stampede the herd toward him. At the last moment he would hide in a crevice while the thundering herd plunged over the precipice to their doom. Scholars believe that Lewis and Clark's party were probably the only white men to actually see the results of this primitive and wasteful method of hunting.

May 29th, 1805 [Lewis] Today we passed on the Stard. side the remains of a vast many mangled carcases of Buffalow which had been driven over a precipice of 120 feet by the Indians and perished; the water appeared to have washed away a part of this immence pile of slaughter and still their remained the fragments of at least a hundred carcases they created a most horrid stench.

You can shorten the trip by several days by taking out at the mouth of the Judith River, where a 40-mile, mostly gravel road leads back to the highway. The river was named after Julia Hancock of Virginia, who would later become William Clark's wife.

May 29th, 1805 [Lewis] passed a handsome river which discharged itself on the Lard. side, Cap. C. thought proper to call it Judieths River.

The weather was still cold, it was raining, and the Corps of Discovery had no wood for a fire. They seemed to make do, however:

May 29th, 1805 [Lewis] accordingly [we] fixed our camp and gave each man a small dram. notwithstanding the allowance of sperits we issued did not exceed one half gill pr. man several of them were considerably effected by it; such is the effects of abstaining for some time from the uce of speritous liquors; they were all very merry.

Below the Judith is the spot where most of the early artists painted their pictures of the steamboats. **Dauphine Rapids** was a major obstacle, especially after midsummer, when the river was low. Often the only way over the bar was to off-load the cargo, tow the now-lightened boat upstream, and then laboriously reload it back again. There was plenty of time to climb to the top of the bluff and set up an easel and record the scene.

Twenty miles below Dauphine Rapids is **Bullwhacker Coulee,** where Captain Lewis spent a memorable day.

May 26th, 1805 [Lewis] In the after part of the day I also walked out and ascended the river hills, which I found sufficiently fortiegueing. on arriving to the summit I thought myself well rapaid for my labour; as from this point I beheld the Rocky Mountains for the first time, these points of the Rocky Mountains were covered with snow and the sun shone on it in such manner as to give me the most plain and satisfactory view. while I viewed these mountains I felt a secret pleasure in finding myself so near the head of the heretofore conceived boundless Missouri, but when I reflected on the difficulties which this snowey barrier would most probably throw in my way to the Pacific, and the sufferings and hardships of myself and party in thim, it in some measure counterballanced the joy I had felt in the first moments in which I gazed on them.

Alas, Lewis was mistaken. The mountains he was looking at are the **Bear's Paw,** an isolated range over 100 miles west of the actual Rockies. It's hard to guess from Lewis's writing whether he knew it beforehand, or whether it was this sight that convinced him there would be no easy portage to the Columbia, that they would find no "northwest passage." If the latter is true, there ought to be some sort of commemoration at the top of this lonely bluff, for western man had been convinced of the existence of such a passage since the 16th century. Captain James Cook searched for it, as had Henry Vancouver. The Hudson's Bay Company had sent Alexander McKenzie on two voyages to the Pacific to find it. Jefferson justified the expenditure of money for this expedition in the hope that it would open up a trade route to Cathay.

June 20th, 1803 [Jefferson] Although your route will be along the channel of the Missouri, yet you will endeavour to inform yourself, by inquiry, of the character and extent of the country watered by its branches, and especially on its southern side. The North river, or Rio Bravo, which runs into the gulf of Mexico, and the North River, or Rio Colorado, which runs into the gulf of California, are understood to be the principal streams heading opposite to the waters of the Missouri, and running southwardly. Whether the dividing grounds between the Missouri and them are mountains or flat lands, what are their distance from the Mis-

souri, the character of the intermediate country, and the people inhabiting it, are worthy of particular inquiry.

Lewis had finally proved that Jefferson's hopes, nurtured by the dreams of explorers for 300 years, were not to be. That same day, Captain Clark commented:

May 26th, 1805 [Clark] This Countrey may with propriety I think be termed the Deserts of America, as I do not conceive any part can ever be settled, as it is deficent in water, Timber & too steep to be tilled.

The next sight along the river is lonely Cow Creek, where on Sept. 23, 1877, Chief Joseph and his weary band tried to buy supplies from an army supply depot. When their offer was refused the Nez Perce took what they needed, burned what was left, and traveled on. The band was to see seven more days of freedom. Cow Creek is 20 miles above **James Kipp State Park,** the final take-out point for the "Wild and Scenic" portion of the Missouri River.

Great Falls of the Missouri

Great Falls, Montana's second largest city, lies about 40 miles southwest of Fort Benton. Somewhere along here you cross the 3000-foot contour, which is suprising because it doesn't seem as if you've climbed a thousand feet since leaving Fort Peck. If you are driving across this undulating plain on a clear day, and most days are, you begin to realize that the mountains are starting to surround you. To the southeast is an isloated range called the **Little Belts;** they are 60 miles away, but the air is so clear it certainly doesn't seem that far. Directly to the south, with the highest peak barely poking up over the horizon, are the **Big Belt Mountains,** and off to the east are the long-sought Rockies themselves.

Ryan Dam

When the Corps of Discovery finally got things straightened out at the Marias Lewis went on ahead while Clark supervised the hiding of the red pirogue that they now felt they could do without. Lewis sensed that he would soon come to the great falls that Minnetarees had told him about, and he was right.

June 13, 1805 [Lewis] I had proceded on this course about two miles whin my ears were saluted with the agreeable sound of a fall of water and advancing a little further I saw the spray arrise above the plain like a collumn of smoke. I did not however loose my direction to this point which soon began to make a roaring too tremendious to be mistaken for any cause short of the great falls of the Missouri.

Incredibly, Lewis was seven miles away from the first of the five cataracts that make up the **Great Falls** of the Missouri when he heard that roar. It's good to keep in mind that in 1805, of all of America's greatest falls—Yellowstone, Yosemite, and Niagara—only the latter had been discovered, and Lewis only knew about that from stories he had heard. He must have had much the same feelings as Dr. David Livingston did 50 years later when he first cast eyes on Africa's Victoria Falls. This was an awesome sight, and Lewis was dumbfounded. To stand where he stood, turn south on a narrow but paved road that leaves U.S. 83 at milepost 6. The 9-mile road leads to the **Ryan Recreation Area,** a heavily forested picnic spot directly below Great Falls. A pedestrian bridge leads to the lovely island where, later that day, Lewis composed some of the most eloquent prose in his entire journal. The Montana Power Company's Ryan Dam, located just above the falls, has destroyed much of what Lewis saw, but if you're lucky enough to be here during a large spring runoff, you will certainly capture some of his excitement.

June 13, 1805 [Lewis] immediately at the cascade the river is about 300 yds. wide, about ninty or a hundred yards of this next the [Left] bluff is a smoth even sheet of water falling over a precipice of at least eighty feet, the remaining part of about 200 yards on my right formes the grandest sight I ever beheld, the irregular and somewhat projecting rocks below receives the water in it's passage down and brakes it into a perfect white foam which assumes a thousand forms in a moment sometimes flying up in jets of sparkling foam to the hight of fifteen or twenty feet and are scarcely formed before large roling bodies of the same beaten and foaming water is thrown over and conceals them. in short the rocks seem to be most happily fixed to present a sheet of the whitest beaten froath. the water after decending strikes against the butment on which I stand and seems to reverberate and being met by the more impetuous courant they roll and swell into half formed billows of great hight which rise and again disappear in an instant.

The next day Lewis explored upriver and found four more major cataracts. They would not have an easy, one-day portage, as the Minnetarees had led him to believe. He didn't know it then, but it would take the Corps of Discovery a full month of backbreaking labor to pass this barrier. Lewis retreated to a point about 4 miles below present-day Ryan Dam. There he charted a course up a creek on the river's south side that led to a more gently rolling plain. He then set the men to building a wagon fashioned of cottonwood trees so they could haul their goods around this great obstacle. Meanwhile, Captain Clark had discovered a sulfur spring that proved to be of immense value, as Sacagawea was becoming deathly sick. Her loss might have scuttled the whole venture, for her help with the Shoshonis was desperately needed.

June 10th, 1805 [Clark] Sahcahgagwea our Indian woman verry sick I blead her.

June 14th, 1805 [Clark] a fine morning the Indian woman complaining all night & excessively bad this morning, her case is somewhat dangerous.

June 16th, 1805 [Lewis] one of the small canoes was left below this rappid in order to pass and repass the river for the purpose of hunting as well as to procure the water of the Sulpher spring, the virtues of wich I now resolved to try on the Indian women. I found that two dozes of barks and opium which I had give here since my arrival had produced an alteration in her pulse for the better; I caused her to drink the mineral water altogether. she complains principally of the lower region of the abdomen, I therefore continued the cataplasms of barks and laudnumn which had been previously used by my friend Capt. Clark.

The next day Lewis happily wrote that "the Indian woman much better today."

The portage route goes up a narrow canyon that bisects the precipitious south side of the river, and then crosses into present-day Malmstrom Air Force Base. The trek covered 18 miles, finally rejoining the river at the southwestern edge of the present-day city of Great Falls. As part of the city's centennial celebration in 1984, a group of 24 local men decided to reenact the portage, keeping as close to the route as possible and using only the tools of the day. They had a time of it. Their moccasins quickly wore out, the prickly pears that so annoyed the Corps of Discovery proved as troublesome as ever, and the wagon, with its wheels made of cottonwood rounds, was exceedingly tiresome to pull. Like the Corps of Discovery, they drank their last drop of whiskey halfway through the portage, and from then on were hot, dry, and no doubt of a cranky nature.

There is no real way for you to follow the portage route; most of it is on private land. It is easy to imagine what the country looked like, though. On the way back to the main highway, park your car on one of the side roads halfway to the top of the bluff, and walk out into the brush. You'll find the prickly pear nestled amid the bunch grass, and if you try and pick this cactuslike shrub, you will learn exactly why the Corps of Discovery hated it so much. A close inspection of this arid land reveals all sorts of surprises. In the spring the thistle and yucca are in bloom; lovely orange flowers poke up out of the sand, and the smell of the sage is overpowering.

Great Falls

For a long time Great Falls was Montana's largest city, thanks in part to having an abundance of cheap hydroelectric power from Ryan and other nearby dams. Anaconda Copper built a huge reduction plant on a bluff overlooking the river. It's gone now; both Anaconda and the copper industry in general have recently seen hard times. Nevertheless, Great Falls is still known as "the electric city" and the local semipro baseball club is the Great Falls Electrics. The city, population 65,000, seems to be clinging to its vitality because of the nearby air base and the fact that it's become the western headquarters for grain companies like General Mills. The local newspaper is the best in the state. The ten-block downtown has seen a few retail businesses flee to outlying shopping districts, but for the most part it is still active. One activity that is gone is the Milwaukee Railroad, which went into bankruptcy a decade ago. The old depot, one of the handsomest buildings in town, is now a restaurant and bar.

Pride of Great Falls is its assocation with the great cowboy and artist, Charles M. Russell. He lived here after he came off the range, and his house and studio are open to the public. A lovely new **art museum,** huge for a town this size, has been built next door. Many of his paintings are on display. Although he was a native of New York, Charlie Russell became Montana's favorite son; his likeness graces the hall of the Capitol in Washington, DC, where the Congress has invited each state to display two statues of their greatest heroes. The **Russell Museum** houses a number of Indian artifacts and includes works by other well-known western artists such as George Catlin and the noted painter and sculptor, Frederic Remington.

Great Falls is a city of tree-lined residential streets and dozens of parks. Almost the entire riverfront is public land, with picnic tables and playing fields overlooking the placid water. A well-used marina is across from a place Lewis and Clark called White Bear Island. The Missouri makes a sharp bend to the right, thus forming both the western and northern boundaries of the city. It's a lovely drive along the bank. On a high bluff, you have a fine view of the uppermost falls and the old smelter site across the river. In a strange way, our heroes were responsible for naming these falls and the town across the way. The Minnetarees had told them that when they came upon an eagle's nest on an island in the river, they would know they had found the true great falls.

June 14th, 1805 [Biddle] Just below the falls is a little island in the middle of the river. Here on a cottonwood tree an eagle had fixed her nest, and seemed the undisputed mistress of a spot, to contest whose dominion neither man nor beast would venture across the gulfs that surround it, and which is further secured by the mist rising from the falls. This solitary bird could not escape the observation of the Indians, who made the eagle's nest a part of their description of the falls, which now proves to be correct in almost every particular, except that they did not do justice to the height.

The name **Black Eagle** has been used to define this spot ever since. The road along the river continues on to a state park and picnic area, where the explorers discovered a giant spring. On this and several other occasions, they used the rather quaint term "fountain" rather than "spring."

June 29th, 1805 [Biddle] they proceeded to the fountain, which is perhaps the largest in America. It is situated in a pleasant level plain, about 25 yards from the river, into which it falls. The water boils up from among the rocks with such force, near the center, that the surface seems higher there than the earth on the sides of the fountain, which is a handsome turf of fine green grass. The water is extremely pure, cold, and pleasant to the taste. It is perfectly transparent.

That's exactly what is looks like today. We now know that Giant Springs is the largest fresh water spring in the world with 388 million gallons of water flowing into the Missouri every single day. The road continues on for a bit and then turns up onto the bluff where the city has built several observation platforms that look out over two of the three other falls. The remnants of Rainbow and Crooked Falls can be seen, but the one named after the Corps of Discovery's John Colter is submerged now.

A number of weird, wild, and scary things happened to the party during the month they spent here. A sudden storm caught Clark, York, Charbonneau, Sacagawea, and her baby in a gully that quickly became a torrent. Charbonneau again panicked, but Clark, now waist-deep in water, managed to push Sacagawea to safety. He lost his knapsack, gun, tomahawk, umbrella, shot pouch, and powder horn, but the most valuable item, a compass, was recovered the next morning. Lewis seemed to be constantly stalked by strange animals, and both captains heard noises they could not explain. It was a wonder that no one was bitten by rattlesnakes, which were abundant everywhere between here and central Oregon. Clark estimated the buffalo population at 10,000. Lewis had to jump into the river to escape one grizzly and Drouillard killed the biggest one they had ever seen. When an iron-framed boat Lewis had designed sank because they had no pine tar to seal the seams of the buffalo hides the men set to work making dugout canoes. The captains set their men to work jerking buffalo, a wise decision, for although they didn't know it at the time, they were about to leave their land of plenty.

The Fourth of July came and went, their second on the trek, and they celebrated by drinking their last drop of spirits. That day Lewis wrote:

July 4th, 1805 [Lewis] have concluded not to dispatch a canoe with a part of our men to St. Louis as we had intended early in the spring. all appear perfectly to have made up their minds to suceed in the expedition or purish in the attempt. we all beleive that we are now about to enter on the most perilous and difficult part of our voyage, yet I see no one repining; all appear ready to meet those difficulties which await us with resolution and becoming fortitude.

Irony and whiskey seemed to go hand in hand with the Corps of Discovery. In Missouri a soldier broke into the whiskey barrel at the site of a future distillery. Today, Montana's only distillery produces Lewis and Clark bourbon, gin, vodka, and schnapps at the exact spot where they drank their last drop.

The party set out again on July 15. They had already used up half the traveling season, and they had no idea how much farther they had to go. If they had had horses, the best thing would have been to strike west, taking an overland route paralleling the Sun River, which would take them to a low pass over the Continental Divide and then to an easy trail leading to a place they would later name Travelers Rest. Lewis and half the party came home this way, learning that they could have shortened their trip by two months. But, of course, they didn't have horses, so they stuck to the Missouri, struggling against its increasingly rapid current. That decision added 400 miles to their trek. State Highway 200 more or less parallels Lewis's homeward trip, so if you, too, want a short cut, take that road. We'll cover that route in "Bitterroots." In the meantime, however, let's continue following Lewis and Clark as they seek out the unknown.

Gates of the Rocky Mountains

For the first time since leaving South Dakota, you find yourself on an interstate. I 15 is the major north-south highway serving the middle-far west. Fortunately, after driving 7 miles you can get off the interstate and follow the old road that, for the next 50 miles, more nearly parallels the Missouri. It soon becomes apparent that the Corps of Discovery finally was entering some real mountains.

July 18th, 1805 [Lewis] saw a large herd of the Bighorned anamals on the immencely high and nearly perpendicular clift opposite us; on the fase of this clift they walked about and bounded from rock to rock with apparent unconcern where it appeared to me that no quadruped could have stood, and from which had they made one false step they must have been precipitated at least a 500 feet. we passed the entrance of a considerable river on the Stard. side. it appears as if it might be navigated but to what extent must be conjectural. this handsome bold and clear stream we named in honor of the Secretary of war calling it Dearborn's river.

Lewis was repaying a favor. The influential secretary, for whom the Michigan city also was named, gave Lewis more or less carte blanche at the army supply depot at Harper's Ferry.

After a bit you have to rejoin the interstate, which leaves the Missouri Valley and goes up the canyon the captains named for John Colter, but is now called Prickly Pear Creek. Sixteen miles north of Helena, turn off at the exit marked "Gates of the Mountains." The road leads down to a reservoir where, during the summer, you can board a boat and see almost exactly what Meriwether Lewis saw in 1805.

July 19th, 1805 [Lewis] when ever we get a view of the lofty summits of the mountains the snow presents itself, altho' we are almost suffocated in this confined valley with heat. this evening we entered the most remarkable clifts that we have yet seen. these clifts rise from the waters edge on either side perpendicularly to the hight of 1200 feet. the river appears to have forced it's way through this immence body of solid rock for the distance of 5¾ Miles. I entered this place and was obliged to continue my rout untill sometime after dark before I found a place sufficiently large to encamp my small party; from the singular appearance of this place I called it the gates of the rocky mountains.

Lewis was the first, but not the only one, to describe the beauties of this spectacular canyon. The painter George Catlin wrote: "In many places . . . there is one continued appearance . . . of some ancient and boundless city in ruins—ramparts, terraces, domes, towers, citadels and castles may be seen—cupolas, and magnificent porticoes, and here and there a solitary column and crumbling pedestal, and even spires of clay which stand alone."

During the summer boats leave hourly for a trip through the canyon, which takes about 2 hours. The captain points out all the sights and explains in geological terms how the defile was formed. There is a chance of seeing a bighorn or a mountain goat, but for that experience it's better to spend some time here. Backpackers take the boat to **Meriwether Campground,** where a trail leads off into the **Gates of the Mountains Wilderness Area**. The local people have learned the fun of packing a picnic lunch, taking an early boat, spending the day in the gorge, and then catching the 5 o'clock boat home.

Helena

Shortly after leaving the Gates of the Mountains, you enter the wide, semiarid Helena basin, which would be one vast desert, were it not for the irrigation projects that have made everything green. Again, it doesn't seem as if you are climbing, but just before you reach the Helena city limits you cross the 4000-foot contour. The state capitol, with its blackened dome, can be seen from a good many miles off.

Helena can rightly be called a tourist town, the first we have encountered since leaving Missouri. During the summer elephant trains leave on an hour's tour of the business and residential districts. Architecturally, this is, with the possible exception of Portland, the most interesting city on our entire journey. The place is blessed with hundreds of lovely turn-of-the-century buildings that miraculously survived a devastating earthquake in 1935.

The story of Helena's founding is well known. In 1864 four itinerant prospectors were returning from Idaho, broke, discouraged, and anxious to get home to Georgia. They decided to take one last chance at finding color in a lonely gulch leading down to Prickly Pear Creek. It wasn't long before Last Chance Gulch became the most-talked-about place in the west. Three years later Helena boasted 45 grocery stores, 5 banks, and 14 saloons. By 1888 the town had 50 millionaires, but had there not been a falling out by two friends from nearby Butte City, Helena would be a ghost town today. The feud was between two copper kings, Marcus Daly and William A. Clark. We'll learn more about these two gentlemen later, but suffice it to say here that when it came time to pick a site for the state capital, whatever town one of them was for, the other was automatically against. Daly, who wanted the capital in his town, **Anaconda,** denied that he spent a million dollars buying votes, but no one believed him. It would have been a good investment if it had worked; he owned all the lots in town. Nobody knows how much Clark spent pushing Helena's cause, but he carried the day. The score: Helena, 27,028; Anaconda, 25,118. Today, two thirds of the working population of Helena works for the state. Since the other third must provide the goods and services for the state employees, it is obvious that if Daly had had his way, Helena would now look much like Montana's first territorial captial, **Bannack**. Weeds grow in Bannack's unpaved streets—its population: 5.

Twenty-five thousand people live here; double that if you count the surrounding countryside. Originally, so the story goes, the name was pronounced with the accent on the second syllable, but to the miners this place was hell, so that's how Helena is pronounced today. Last Chance Gulch became Main Street, which has been graciously restored as a pedestrian mall. The site where the first pan of gold was found is under the stately **Montana Club,** located a block away. The Northwestern Bank of Helena has a **gold museum** featuring a two-pound nugget found nearby. Also shown is a pan of gold washed from the dirt excavated for the bank's foundation in 1959. The principal downtown landmark is a 100-year-old wooden fire tower. Looming above the fire tower is mile-high **Mt. Helena,** where several nature trails lead to the summit.

Helena is full of buildings you wouldn't expect to see in a town this size. Many of the homes the millionaires built are still in use, though one wonders who can afford the heating bills. The stately brick **"Old Governors Mansion"** is now a museum. An old sandstone church now houses a legitimate theater that puts on year-round performances. The **City Hall,** a Byzantine structure with a single minaret, at one time was a Shriner's temple. The huge **Cathedral of St. Helena** is reputedly a partial replica of the Cathedral in Milan, though anyone who has seen both will have a bit of trouble discerning the similarity. The Catholic **Carroll College,** which sits atop a hill, is so prominent that tourists often confuse it with the capitol.

A tour of the **capitol building** is a rewarding experience. The main floor is graced by a statue of Jeanette Rankin, the first woman ever to be elected to the United States Congress. A lifelong pacifist, she voted against our entry into World War I, lost her seat because of it, and then won it back just before Pearl Harbor. Her convictions were deep-rooted; she cast the only nay vote on entry into World War II. Highlight of the capitol tour is Charlie Russell's masterpiece above the speaker's desk in the House of Representatives' Chamber. Titled "Lewis and Clark Meeting the Flathead Indians at Ross' Hole," it shows our heroes near Missoula on their return from the Pacific. Another mural, outside the chamber, shows Lewis and Clark at Three Forks, accompanied by Sacagawea (pointing the way, of course), York, and two other men, presumably Drouillard and Charbonneau.

Across the street from the capitol is an absolutely first-class **museum and art gallery**. One room features paintings by Charlie Russell and his contemporaries, another the pioneer photographs of F. Jay Haynes, who recorded some memorable scenes of early Montana life.

Three Forks

The city of Helena is about 10 miles from the Missouri, so the Corps of Discovery never actually saw the site. By this time they were heading due south, in effect losing what they had gained the previous summer. They seemed to be moving away from, not toward, the high mountains. The captains were on the lookout for one other landmark the Minnetarees had told them about. They were also getting anxious.

July 22nd, 1805 [Lewis] The Indian woman recognizes the country and assures us that this is the river on which her relations live, and that the three forks are at no great distance. this piece of information has cheered the sperits of the party who now begin to console themselves with the anticipation of shortly seeing the head of the missouri yet unknown to the civilized world.

July 24th, 1805 [Lewis] I fear every day that we shall meet with some considerable falls or obstruction in the river notwithstanding the information of the Indian woman to the contrary. I can scarcely form an idea of a river runing to great extent through such a rough mountainous country without having it's stream intersepted by some difficult and dangerous rappids or falls.

July 27th, 1805 [Lewis] we begin to feel considerable anxiety with rispect to the Snake Indians. if we do not find them or some other nation who have horses I fear the successful issue of our voyage will be very doubtful we are now several hundred miles within the bosom of this wild and mountainous country, where game may rationally be expected shortly to become scarce and subsistence precarious without any information with rispect to the country not knowing how far these mountains continue, or wher to direct our course to pass them to advantage or inerrsept a navigable branch of the columbia, however I still hope for the best.

Lewis's choice of the word bosom is an interesting one; it suits this country exactly. The Rocky Mountains here are not a single rampart like they are farther north in Glacier National Park, but instead are a dozen isolated ranges, separated by wide fertile plains. The Corps of Discovery, following the Missouri River, entered one of these plains immediately after emerging from the Gates of the Mountains. After passing through another small gap, the explorers are now in the second valley, and in due time they will enter a third. Each plain is successively higher, the grass seems greener, and the air purer. This is the beginning of Montana's justly famous resort country.

One wonders what Lewis's anxiety level would have been if he had known that it would be another month before he would have his horses. One also wonders what was going through the minds of three of the crew. John Potts, John Colter, and George Drouillard would return to this spot again. One would barely escape with his life, and the other two didn't have a chance.

Captain Clark, who had gone on ahead, got to the Three Forks first.

July 25th, 1805 [Clark] a fine morning we proceeded on a fiew miles to the three forks of the Missouri those three forks are nearly of a Size, the North fork appears to have the most water and must be Considered as the one best calculated for us to assend. The bottoms are extencive and tolerable land covered with tall grass & prickley pears. The hills & mountains are high Steep & rockey. The river verry much divided by Islands. Musquetors verry troublesome untill the Mountain breeze sprung up, which was a little after night.

Many things have changed since Lewis and Clark passed this way, but one thing hasn't. The "Musquetors are verry troublesome"! Sixty miles south of Helena on U.S. Highway 12, you come to Interstate 90. Turn left 4 miles to the **Missouri Headwaters State Park** exit, and then drive north about a mile to an isolated bluff on your right. Now, before you do anything else, before you even open the car door, douse yourself with a liberal coating of mosquito repellent. This place is the Times Square of the mosquito crowd.

Lewis arrived at Three Forks two days later and promptly climbed a bluff to have a look around.

July 27th, 1805 [Lewis] the country opens suddonly to extensive and beatifull plains and meadows which apper to be surrounded in every direction with distant and lofty mountains; I commandanded a most perfect view of the neighbouring country. from E. to S. between the S.E. and middle forks a distant range of lofty mountains ran their snow clad tops above the irregular and broken mountains which lie adjacent to this beautiful spot. between the middle and S.E. forks near their junction with the S.W. fork there is a handsome site for a fortification.

You're now exactly 2475 miles above the Mississippi River. The U.S. Geological Survey identifies this as the source of the Missouri, for each of the three upstream rivers carries a new name.

July 28th, 1805 [Lewis] we called the S.W. fork Jefferson's River in honor of that illustrious personage Thomas Jefferson. the Middle fork we called Madison's River in honor of James Madison [Secretary of State at the time], and the S.E. Fork we called Gallitin's River in honor of Albert Gallitin [Secretary of the Treasury].

This was beaver country, and although Lewis thought them a delicacy, their presence wasn't making travel very easy.

> *July 30th, 1805 [Lewis] saw a vast number of beaver in many large dams which they had maid. I directed my course to the high plain to the right which I gained after some time with much difficulty and waiding many beaver dams to my waist in mud and water. I would willingly have joined the canoes but the brush were so thick, the river crooked and bottoms intercepted in such a manner by the beaver dams, that I found it uceless to find them.*

The men of the Corps of Discovery would return to St. Louis with stories of the fantastic beaver hunting at Three Forks. For the next six years numerous trappers would come here, certain they would get rich. What they didn't seem to understand was that this was kind of a no-man's land, a place where the Blackfeet, Assiniboines, and Crows raided the Shoshonis and Flatheads when the latter ventured eastward over the mountains to hunt for buffalo. It's one of those landmarks that everyone, Indian and white, knew about, and thus became a rendezvous for trapper and trader, marauder and thief. Today, Three Forks seems to be in the middle of nowhere, a surprising thing for such an important spot. Captain Clark also noted that it was "a position admirably well calculated for a fort." The present-day loneliness becomes understandable, though, when you consider what went on here, both before and after Lewis and Clark passed by.

> *July 28th, 1805 [Lewis] Our present camp is precisely on the spot that the Snake Indians were encamped at the time the Minnetares. first came in sight of them five years since. the Minnetares pursued, attacked them, killed 4 men and 4 women a number of boys and made prisoners of all the females and four boys,* Sah-cah-gar-we-ah *our Indian woman was one of the female prisoners taken at that time; tho' I cannot discover that she shews any immotion of sorrow in recollecting this event, or of joy in being again restored to her native country; if she has enough to eat and a few trinkets to wear I believe she would be perfectly content anywhere.*

Perhaps her lack of joy was because she had a premonition of what was to happen to her comrades a couple of years later. In 1808 John Colter, fresh from his epic trip through present-day Yellowstone Park, was wounded a few miles from here in a skirmish between the Crows and Flatheads. A few months later, when Colter was trapping with John Potts, that pair was surprised by a band of Blackfeet who immediately became angry when they recognized Colter as being a friend of the hated Crows. Potts put up a fight and was killed. The Blackfoot chief decided that Colter could give his young warriors some sport. Stripped of his clothes and moccasins and given a 100-yard head start, Colter was ordered to run for his life across the cactus-studded plain. Somehow, bleeding from nose and mouth, he managed to out-distance all but one warrior, whom he suddenly turned on. The startled warrior tried to stop, tripped, and lost his spear. Colter seized the weapon, slew the brave, and ran some more. Five miles later he came to the river, dove under a pile of driftwood, and hid. Luck was on his side; from his lair he watched his pursuers search for him and on several occasions walk over the very entanglement under which he was hiding. When night came he swam downstream. Seven days later, exhausted and almost starved, he stumbled into Manuel Lisa's fort on the Big Horn, 200 miles away.

Five years after the Voyage of Discovery passed by a party of 32 trappers under the command of Lisa's lieutenant, Pierre Menard, built a fort just where the captains had suggested. A week later, who should show up but John Colter, fresh from another skirmish with the Blackfeet. Nothing seemed to daunt this fearless mountain man, but when five of Menard's men were killed and mutilated he evidently decided he had exhausted his luck; he headed back to Missouri, never to return to the mountains again. That same year George Drouillard arrived back on the scene. He wasn't so lucky. He and two Shawnee companions died in an ambush while trapping along the Jefferson. His decapitated and mutilated body was buried somewhere nearby. Pierre Menard had had enough—he retired from the scene before the advent of winter. It wasn't the end of interest in the area, though. Before the beaver were trapped almost to extinction in 1840, some of the most famous of all the early mountain men frequented the area: Jim Bridger, Kitt Carson, James Beckworth, and Jedediah Smith. A town, **Gallatin City,** was established in 1865, but it didn't last long either. It wasn't until the coming of the railroad after the turn of the century that any permanent settlement was established at Three Forks.

The State of Montana has done a fine job with the **Missouri Head-**

waters Park. Two Gallatin City buildings have been preserved and an outdoor visitors center describes some of the events noted above. Many of the indigenous plants are identified, including buckwheat bush, sage, pin-cushion cactus, prickly pear, blue bunch, wheat grass, and salt grass. The highlight of your visit will probably be a stroll to the top of the bluff—the same one Lewis recommended for a fort—where you get a sweeping panorama of the four rivers and the surrounding countryside. The lovely **Tobacco Root Mountains** are prominent on the southwestern skyline.

The town of Three Forks is located about three miles away. For most of the year it's a sleepy little place, but when the salmon-fly hatch is on, sportsmen from all over the country flock to this anglers' paradise. Many fly-fishermen believe the nearby streams are the best in the world, though perhaps not quite as good as they were in 1805. One day Captain Lewis set the men to work making what he called a bush drag. Within two hours "they caught 528 very good fish, most of them large trout." Only about 1200 people live in Three Forks, but the town boasts a dandy old-fashioned hotel, named for the maiden who was kidnapped here. Staying in the **Sacajawea Inn** is a great experience if you don't mind sharing a bathroom. The operator of the local archery store must have both a sense of humor and an interest in architecture. His shop is called the "Bow Haus."

THE YELLOWSTONE RIVER

On their return trip to St. Louis, Lewis and Clark decided to split up the group so they could explore some of what they missed on their westward trek. Captain Clark, with 21 of the party, took a southern route, arriving at Three Forks, Montana, in mid-July. Here he again split his group, with Sergeant Ordway taking nine men in six canoes down the Missouri, following the same route the explorers had used the previous summer. Clark, with the remaining ten, plus Sacagawea and her child, proceeded by land across a pass and into the valley of the Yellowstone River, where they built canoes and floated down to its confluence with the Missouri. We pick up their trail at **Fort Union** on the North Dakota/Montana border. Clark's party arrived in early August. Lewis, coming down from Great Falls, Montana, got here four days later.

August 3rd, 1806 [Clark] last night the Musquetors was so troublesom that no one of the party Slept half the night. on the side of [a] bluff I saw some of the Mountain Bighorn animals. I assended the hill below the Bluff. The Musquetors wer so noumerrous that I could not shute with any certainty and therefore soon returned to the Canoes. at 8. A.M. I arrived at the junction of the Rochejhone with the Missouri, and formed my camp imediately.

Sidney/Glendive

In order to avoid having to backtrack into Williston, take the dirt road west from Fort Union for a few miles, and then turn south on another unpaved road that crosses the Missouri on a new bridge. You then have about ten more miles of gravel road to follow before reaching Montana Highway 200. The first Montana town you come to is Sidney. This agricultural center, with a population of 6000, has recently built the **MonDak Heritage Center,** which houses a museum and art gallery. Fifty miles farther up the river is Glendive, which, legend has it, was named for a frontier saloon owned by a fellow named Glen. Apparently it wasn't much of a place. The trappers and hunters who came here for the rotgut Glen served called it Glen's dive. The Northern Pacific Railroad arrived here in 1881, and the town quickly became a shipping center for the nearby ranches. Six thousand people now live in Glendive. A collection of early prairie artifacts is housed in the **Frontier Gateway Museum**.

The lower Yellowstone River is quite unlike the managed and tailored Missouri. Here the river flows free, as it has all the way from its source in Yellowstone Park. The entire Yellowstone watershed has only three dams, all on a tributary river, the Big Horn. Between Glendive and Sydney, the Yellowstone is called a braided river because the constantly changing channels weave in and out of each other, creating a thousand sloughs and a million cottonwood-forested islands. An estimated 6000 white-tailed deer call these islands home, as do countless Canada geese, bald and golden eagles, and pelicans. A great blue heron rookery contains 50 nests, each as large as a laundry basket. The locals boast that the fishing on the lower river is as challenging as anywhere, with walleye, northern pike, ling, and catfish being the principal catch. The river also contains the prehistoric-looking paddlefish, which ranges from 50 to a 150 pounds. The free-flowing river is, of course, popular for canoeists and rafters.

The cattle industry in Montana more or less got its start near here in 1866, when one Nelson Story drove 600 head of longhorns up from Texas. The trappers and mountain men, used to a diet of buffalo, didn't particularly like the meat; it was too sweet. Nevertheless, by 1877 cattlemen, who never owned more than a few hundred acres, were grazing 20,000 head or more. They had no money, so what land they needed they just took—from the Crow, the public domain, and from each other. Rustling was rampant for a while, leading to the inevitable organization of vigilante committees. By 1880 Montana probably had 48,000 horses, 1600 mules, 250,000 sheep, and 275,000 head of cattle. Three years later 600,000 head of cattle and 500,000 sheep were causing serious overgrazing problems on the same land that had previously supported millions of buffalo.

Then came "The Hard Winter." There was a good rain in 1885, but a dry summer the next year caused numerous range fires. The winter of 1885–1886 was mild, but the next summer again turned hot. A bad winter to the south forced an additional 200,000 head into the area. Then a sudden drop in prices caused cattlemen to withhold their herds from market, setting the stage for the disaster that followed. The fall and early winter of 1886–1887 saw no rain; the grass was thin. It snowed most of November, then thawed in mid-December. Then very low temperatures and bitter winds created an ice sheet. The cattle lacerated their noses and legs trying to get to what grass there was. The temperature in Glendive during the first two weeks in Feburary *averaged 27.5° below zero*. March brought a "chinook," the thawing wind that often proceeds spring, but by then it was too late—60–80% of the herds were lost. Charlie Russell was working at the time as a cowhand in the Judith Basin, about 15 miles west of here. His painting of a forlorn-looking heifer, titled "Waiting for a Chinook," launched his artistic career. A new industry emerged from the disaster. Cattlemen could no longer be nomads. Barbed wire replaced the branding iron, and hay, when needed, was brought in by railroad. The sheep industry did better for a while, with six million grazing in the Montana plains in 1900. But the woolen mills were back east, and so ten years later that boom went bust. It was time for the sodbuster to take over.

Four principal rivers enter the Yellowstone from the south. The first you come to is the Powder, which the party passed on the last day of July. Clark's journal notes the great quantity of coal found in the area. Recently, partly because of the energy crisis, the Powder River basin has seen a great deal of coal activity.

Miles City

The second major river, the Tongue, joins the Yellowstone at Miles City, a town of 10,000 in the heart of the cattle country. The place was named after Colonel Nelson (Bear Coat) Miles, commandant of nearby Fort Keogh, who had his day of glory when he received Chief Joseph's surrender at the Bear's Paw Mountains. Miles City is a nice western town with a number of good motels and restaurants. At one time, however, it was described as "a miserable little place; the backwash of western society, where the chief occupation was selling booze to the soldiers and the Indians." The principal attraction, the **Range Riders Museum,** has exhibits and memorabilia of the days of the open range. Two miles south of town is the site of **Fort Keogh,** at one time the largest army post in Montana. Built after the Battle of Little Big Horn, the post continued until the close of the so-called Indian Wars in 1908.

Big Horn River

Seventy-five miles west of Miles City you come to a spot where virtually nothing exists today, but at one time this was the cradle of white civilization in upper "Louisiana." Along this stretch of road you get your first glimpse of the Rocky Mountains.

> *July 26th, 1806 [Clark] fortunately for us we found an excellent chanel to pass down on the right of a Stoney Island half a Mile below this bad place, we arived at the enterance of Big Horn River. I am informed by the Menetarres Indians and others that this River takes its rise in the Rocky mountains with the heads of the river plate and at no great distance from the river Rochejhone. the river is said to abound in beaver it is inhabited by a great number of roveing Indians of the Crow Nation.*

Clark had his geography exactly right. The Platte, Yellowstone, and Big Horn rivers all have their sources in the southeastern corner of Yellowstone National Park.

> *July 27th, 1806 [Clark] I marked my name with red paint on a cotton tree near my camp, and Set out at an early hour. The Buffalow and Elk is estonishingly noumerous, particularly the Elk which are so jintle that we frequently pass within 20 or 30 paces of them without their being the least alarmd. when we pass the Big horn I take my leave of the View of the tremendious chain of Rocky Mountains white with Snow in View of which I have been since the 1st of May last.*

The confluence of the Yellowstone and Big Horn rivers is the site of the first American settlement in the upper Missouri basin. Manuel Lisa, having learned from Lewis and Clark about the riches to be had trading with the native tribes, formed a partnership with Pierre Menard and William Morrison, both of whom were fellow merchants from St. Louis. With the help of the great Drouillard, Lisa recruited about 50 men, including our friends John Potts and Sergeant Prior. They set out upriver in two keel boats in the spring of 1807. Along the way, who should they encounter but John Colter, returning alone to St. Louis. Colter never was a city boy; the lure of the mountains was irresistible, so for the second time in two years he was persuaded to turn back. The party was to have a bit of difficulty along the way, particularly with the Arikaras, but thanks to Lisa's firmness, they managed to make their way up the Missouri and Yellowstone rivers unscathed. However, they arrived here too late in the season to set their traps. The men promptly set about building Fort Raymond. Game was plentiful, though they unwittingly had to share what they shot with the grizzly bears.

Lisa's objective was to seek out the native tribes, particularly the

Crow and Cheyenne, and to induce them to come in and trade. To this end he dispatched John Colter on his famous solo journey into the wilderness. Provisioned with 35 pounds of ammunition, clothes, and trade goods, and blessed with at least one good horse, Colter set out on a trip that would take him south along the Big Horn, then west up the Wind River and down into present-day Jackson's Hole, where he became the first white man to set eyes on Les Trois Tetons. He then skirted around the Tetons, dropped into Pierre's Hole, a latter-day fur rendezvous on a tributary of the Colorado River, headed north, where he skirted Yellowstone Lake, and finally returned to Fort Raymond via the Clark's Fork of the Yellowstone. No one believed the stories he had to tell about fumaroles, smoke, mud pots, and hot water spouting up from the ground. What soon became known as Colter's Hell, and later Yellowstone National Park, was simply a figment of this demented man's imagination.

Lisa quit the region in 1811; it proved to be too far into the backcountry. Then the War of 1812 put a kink in all trapping plans. Armed and encouraged by the British, the northern tribes played havoc with any Americans venturing into this country. By the end of the war Lisa had quit the fur business for good. He died in St. Louis in 1820. That same year another effort was made to establish an American presence here. Lisa's old partner, General William H. Ashley, placed a famous ad in the St. Louis *Gazette and Public Advertiser,* which said in part: "Wanted: 100 Enterprising Young Men . . . to ascend the Missouri to its source, here to be employed for one, two, or three years." Ashley, in partnership with Pierre and Auguste Chouteau, Reuben Lewis (Meriwether's brother), William Clark, Sylvester Labadie, and Andrew Henry, formed the Rocky Mountain Fur Company, which was to revolutionize the way peltry was brought to market. Rather than rely on the local tribes to bring in furs, men in groups of twos and threes were sent out to work the streams, avoiding the Indians as much as possible. Many of the men who answered that ad were to become folk heroes in the winning of the west. Jedediah Smith signed up, as did the five Sublette brothers, Robert Cambell, James Clyman, David E. Jackson, Etinenne Provost, and a half-blood blacksmith named James Beckworth. The last factor to occupy the fort here, Joshua Pilcher, was traveling downriver with $25,000 worth of pelts when he and his lieutenants, W. H. Vanderburgh, Robert Jones, and Michael Immel, were attacked by the Blackfeet. Immel and Jones were killed; the other two gave up their cargo and ran. No trace of either fort exists today.

The mouth of the Big Horn River was generally considered the head of navigation for steamboats, but on June 25, 1876, Grant Marsh, captain of the *Far West,* had a special reason to push his craft farther upstream. Scouts had given him word of Col. Custer's fate, and he steamed up the Big Horn to within 15 miles of the battle, where he picked up the survivors of the Seventh Cavalry. His 10-day race back to Fort Lincoln (Bismarck, North Dakota) set a record for that run.

Custer Battlefield

The battlefield is 45 miles south of the Yellowstone River, adjacent to Interstate 90. The National Park Service operates a visitors center where mementos of the fight are displayed and movies explain the events leading up to the disaster. Rangers give lectures on the lawn and a Park Service bus tours the site. There are essentially two battlefields at the monument: the spot where Custer and 265 of his men were massacred, and the area three miles away where Captains Reno and Benteen, Custer's subordinates, were laid under siege. A lovely cemetery adjoins the visitors center.

In touring the site, you learn that Custer was an enigma. Last in his class at West Point, he distinguished himself at Bull Run and Gettysburg, where he earned the brevet rank of brigadier general of the Michigan militia. Having emerged from the war as one of its most brilliant cavalry leaders, he was given the rank of lieutenant colonel in the regular army. He became a public idol and had presidential ambitions. But U. S. Grant, who detested him, kept him assigned to out-of-the-way posts, during which time he wrote a book, *My Life on the Plains,* which his subordinate at the Little Big Horn, Captain Frederick Benteen, called "My Lie on the Plains." The Sioux called him "Yellow Hair"; his troops called him "Iron Butt." The defeat of the Sioux was to be his chance to regain his falling star.

A visit to the monument seems to leave you with more questions than answers. Why weren't more scouts sent out, and why didn't Custer believe the ones that were? Why didn't he wait for General Terry's forces as he was ordered to do? Why did he split his cavalry into three separate groups? Could Reno and Benteen, both of whom disliked their leader intensely, have come to his aid? What is it about this dashing man that so captures our imagination, a century after the tragedy? Why was Custer's scalp left intact while most other bodies were severely mutilated? Hundreds of books have and will be written to try to explain the enigma that was George Armstrong Custer.

One thing is clear. The battle led to a massive arms buildup in the area. Prior to the massacre, 600 soldiers were assigned to police 1300

Indians. By the time of Joseph's surrender at the Bear's Paw, there were 3000 soldiers in the field, and eight additional forts, one of which was named for the dead hero. By the mid-1880s the government was spending $7 a year to support each Indian living on a reservation, and $1000 a year to put a soldier into the field.

Crow Agency

The Custer monument is located on the Crow Reservation. Ironically, none of the 2000–5000 Indians who took part in the battle were Crow. Sitting Bull's warriors were Sioux and Cheyenne. The army considered the Crow to be the most loyal tribe on the prairie, employing many of them as scouts. In Lewis and Clark's time, they had the reputation of being the world's greatest horse thieves. Clark never encountered a Crow face to face, but he knew they were here. His journal entry for July 18 contains his only reference to smoke signals.

July 18th, 1806 [Clark] I observed a Smoke rise in the plains toward the termonation of the rocky mountains. this smoke must be raised by the Crow Indians as a Signal for us, or other bands. I think it most probable that they have discovered our trail and takeing us to be Shoshones, have made this Smoke to Shew where they are—or otherwise takeing us to be their Enemy made this signal for other bands to be on their guard.

July 21st, 1806 [Clark] This morning I was informed that Half of our horses were absent. Sent out Shannon Bratten and Shabono to hunt them. I am apprehensive that the indians have Stolen our horses, and probably those who had made the Smoke a fiew days passed.

He was right. The scouts never found the stolen horses. Later, after canoes were constructed, Clark sent several of his men overland with the remaining stock of horses, which he hoped to use for trade with the Mandans. The Crows took every single horse; the men were forced back onto the river where they rode in bull boats that they had constructed out of buffalo hides.

The headquarters of the Crow tribe is at Crow Agency, known as the "Tipi Capital" of the world. The large trees and grassy parks give the town a friendly feel. An excellent description of life on the plains is given in an oral history book, *Two Leggings: The Making of a Crow Warrior*.

Hardin

Hardin, the commercial center of Big Horn County, is a pleasant town of 3500 people who shop at some beautiful old brick buildings gracing the three-block main street. The county **Historical Museum** is one of the best in the area. Next door is a beautifully refurbished farmhouse, circa 1911, and behind that is an old log cabin, a lovely little church, and a barn full of old-fashioned farm implements. An **art gallery** is located in the old jail house.

Pompeys Pillar

Thirty miles east of Billings, alongside Interstate 94, is Pompeys Pillar, the only spot on the Lewis and Clark trail where there is physical evidence of their having passed this way.

July 25th, 1806 [Clark] at 4 P M arived at a remarkable rock situated in an extensive bottom. this rock I ascended and from its top had a most extensive view in every direction. This rock which I shall call Pompy's Tower is 200 feet high and 400 paces in secumpherance and only axcessable on one Side. The nativs have ingraved on the face of this rock the figures of animals &c near which I marked my name and the day of the month & year. From the top of this Tower I could discover two low Mountains & the Rocky Mts covered with Snow one of them appeared to be extencive.

The "extencive" mountain he described was probably the **Pryor Mountains,** located about 40 miles south of here, which were named after the Corps of Discovery's Sergeant Prior. Captain Clark's graffiti has been carefully preserved by the owners of this land, who now operate it as an outdoor museum. Numerous other names have been scratched into the soft sandstone. Capt. Grant Marsh inscribed "Josephine, June 3, 1875," to mark the visit of one of the other steamboats he commanded. George Custer and several hundred of his men camped here in 1873. While many were swimming in the river they were ambushed by the Sioux. Unlike the events of three years later, no one was injured.

A few miles before you reach Billings, a dirt road leads into the **Bitter Creek valley** where, tucked into a sandstone outcropping, is a cave adorned with Indian pictographs. This major archaeological site has interpretive panels describing how man has utilized this environment to fulfill his basic needs for 5000 years.

Billings

When Captain Clark's party camped near present-day Billings they had some nice things to say about the area.

> *July 24th, 1806 [Clark] the earth which is rich is covered with wild rye and a Species of grass resembling the bluegrass, and a mixture of Sweet grass which the Indian plat and ware around their necks for its cent which is of a strong sent like that of the Vinella. I landed on the Lard Side walked out into the bottom and killed the fatest Buck I every saw; Shields killed a deer and my man york killed a Buffalow Bull.*

The probable place where they camped that night is in the middle of an oil refinery. Your first view of Montana's largest city isn't all that nice, but once you leave the interstate and cross the railroad, you'll be pleasantly surprised at how nice a town Billings seems to be. Sixty-seven thousand people live within the city limits, and 100,000 live in the metropolitan area of this thriving commercial and industrial center. Incredibly, there is not another city larger than Billings within 500 miles in any direction. Billings, named after one of the early presidents of the Northern Pacific Railroad, has the most extensive shopping facilities between Omaha and Portland. **Hart-Albin,** a first-class department store, carries all the latest fashions, yet nearby you can shop at **Connolly's Western Ware,** an emporium that supports an inventory of at least 1000 pairs of cowboy boots and 500 hats. The 19-story **Sheridan Hotel** is the tallest building in the state. The locals call it "the big sky scraper." Across the street, the **Northern,** one of the oldest hotels in Montana, preserves its stately image. A nearby section of old buildings along the railroad tracks is being renovated. Of special interest is the old **Carlin Hotel,** which boasts a theater pipe organ, and the **Rex Hotel,** which was built in 1911 with support from Buffalo Bill Cody. Billings' restaurants range from the elegant **Golden Belle** to the **Spur,** a huge truck-driver-and-cowboy hangout, where the waitresses are tough enough to handle the most rowdy customers.

Billings is the principal medical center for this part of the country, and is home to **Eastern Montana College,** one of the fastest growing institutions in the state. Lovely homes, set on tree-lined streets, surround the campus and medical center. One old brownstone mansion called the **Castle,** built in 1902, is now an art gallery. Several other homes in the neighborhood have been made into small shops. The **Yellowstone Art Center** is one of the best in the state, and another gallery is located at Eastern Montana College. A nostalgic piece of sculpture is located at the entrance to the airport. The noted cowboy movie star William S. Hart posed for the "Range Rider," sculpted by Charles Christadora.

The old public library, located next to the train station, is now the **Western Heritage Center,** with extensive collections of memorabilia describing life on the plains. The **Yellowstone County Museum,** partially housed in a circa 1893 log house, is located near the airport. The collection includes Indian artifacts, a steam locomotive, horse-drawn vehicles, guns, saddles, and farm implements from the early days of homesteading. **Boothill Cemetery,** so named because most "cashed in with their boots on," is located nearby.

Laurel

Fifteen miles west of Billings is the little town of Laurel where you must make a decision. The route of William Clark's return trek follows the Yellowstone River to Livingston, a distance of a hundred miles. It is an interesting drive as we shall see, but you also have an opportunity to take an alternate route through the Bear Tooth Mountains. Let's explore that country first.

Side trip to the Bear Tooth Mountains

Television's vagabond reporter, Charles Kuralt, calls the route over the Bear Tooths "the most scenic drive in America." The 230-mile trip from Billings to Livingston via the Bear Tooth Pass can be made in a day, but it's best to go as slowly as your time permits; there are several places to stay along the way. Begin by taking U.S. Highway 212 south from Laurel to the semiresort town of Red Lodge.

Red Lodge Red Lodge has much the same flavor and character as Aspen, Colorado, did 30 years ago. Like Aspen, it was a mining town before it became a ski resort, though the mines here produced prosaic coal, not exotic silver. Unlike Aspen, Red Lodge has not yet been discovered by America's jet set. Today, Aspen's old brick buildings house gourmet delis and boutiques that sell Pierre Cardin fashions. Red Lodge's century-old structures, by contrast, house merchants like the Capital Dollar Store, Rexall Drug, and Coast to Coast Hardware. Twenty-five hundred people live in this mile-high city. The four-block-long shopping district has a nice, unpretentious, hometown feel; shoppers seem to know each other. During the summer two rodeos are staged here and the whole town turns out for its week-long Festival of Nations, held in August. The ski resort, located 6 miles west of town, has a top elevation of 9400 feet and a vertical drop of 2000 feet.

The 67-mile-long Bear Tooth Highway, which connects Red Lodge with the northeast entrance to Yellowstone Park, was built in 1936 to give the motoring public a chance to visit some really high mountains on their way to see the thermal delights inside the park. The road, which is usually open between Memorial Day and Halloween, plunges into **Rock Creek Canyon,** and immediately you find yourself surrounded by almost vertical granite walls. Some daring switchbacks take you up the south cliff, and within a few miles you come to a belvedere looking out over the canyon that you just left. It's now 4000 feet below you. A few miles farther on the road circles around a hill on a curve, known locally as the Mae West switchback, and then heads for the summit, which, at 10,946 feet, is one of the highest passes in the country. You're on the **Bear Tooth Plateau,** and even in summer snow piles up along the road. The highway remains at this high elevation for nearly 5 miles, crossing into Wyoming as it winds through country that looks much like Norway's Hardangervida. This is arctic tundra. Tiny little flowers survive in a land where no trees can grow. You feel as if you're at the top of the world, which in fact is the name of a small resort a little way down the road.

Granite Peak, Montana's highest at 12,799 feet, looms above the plateau, and a little farther on you get a good view of the Bear's Tooth, the crag that gave this mountain range its name. Off to the west are two more lovely spires, Pilot and Index peaks. The road then drops a bit into glaciated country, filled with hundreds of rock-bound lakes and azure-blue tarns. You're now in the headwaters of **Clark's Fork** of the Yellowstone, one of the finest fishing streams in a state noted for its fishing. The road climbs again and you cross back into Montana over a small pass named for our friend John Colter, who came through here in 1808 on his way home from Jackson's Hole. Chief Joseph and his band of Nez Perce also crossed through here on their flight from the Big Hole.

Cooke City/Silver Gate In the 1880s gold was discovered along Soda Butte Creek, and within a short time the new town of *Shoofly* boasted 135 log cabins, 2 stores, and 13 saloons. Money was needed to work these hard-rock mines. It came from the east. When Jay Cooke, Jr., son of the railroad baron, put up a stake large enough to keep the mines open the locals thought it prudent to give the town his name. The mines never amounted to much, but enough people stayed on to enjoy life in this beautiful valley that the highway department felt obliged to give it a first-class road, which is now open all year. Cooke City and its neighbor, Silver Gate, are low-key resort towns with rustic, log-cabin-type motels and numerous places to eat. A large old hotel in Silver Gate is built entirely of skinned lodgepole pine logs.

Yellowstone Park The highway enters Yellowstone Park at the little-used Northeast entrance, follows Soda Butte Creek to its junction with the Lamar River, and then follows that lovely stream to its confluence with the Yellowstone. From there you leave the river, travel overland to **Mammoth Hot Springs,** and then turn right to the North entrance at Gardiner. A description of Yellowstone Park is beyond the purview of this book. Suffice it to say that there are a host of sights to see and things to do. All reservations for accommodations are handled through the park concessionaire, which operates two facilities along our route. **Roosevelt Lodge** at Tower Junction has rustic cabins; **Mammoth Springs** has hotel and motel rooms. All told, there are 8000 rooms in the park, and at least that many RV sites. Most campgrounds prohibit tent camping because park rangers are worried about the bears.

Sixty years would elapse between John Colter's swing through here and the arrival of the first party that deliberately set out to explore the mysteries of the fumaroles, mud pots, geysers, and soda fountains. Some men from Montana's gold fields tried to explore the Gallatin basin in 1867, but were repulsed by Indians. It was not until 1869 that three men from Helena, David Folsom, C. W. Cook, and a ranch hand named Peterson, made a successful trek up the Yellowstone to its headwaters at Yellowstone Lake. From there they headed west, crossed over the Continental Divide, went past some of the geysers, and headed home by an unknown route. The following year N. P. Langford of Helena convinced the army that it should form a survey party, which it did, and thankfully came to the conclusion that the land should be made into a park. In 1871 Ferdinand Hayden, head of the U.S. Geological Survey, led a civilian trek through the area and confirmed the wisdom of Langford's pleas. Yellowstone became our first national park in 1872.

Gardiner, Montana, is the oldest entrance to Yellowstone, and the only one open all year. Straddling a little-used road is a stone arch, the official gateway to the park, which was dedicated by Teddy Roosevelt in 1903. During the early days park visitors arrived here by train. The Great Northern Railroad built a spur line up the Yellowstone River valley from Livingston, which then connected with stagecoaches bound for Old Faithful Inn. Today, in addition to fishing, one of the big attractions is the whitewater rafting on the Yellowstone River. Several outfitters offer half-day and one-day trips. Gardiner has a few motels and several restaurants that offer a change from the more institutional atmosphere found inside the park.

Chico Hot Springs Fifteen miles below Gardiner, the steep-sided Yellowstone canyon opens out into the wide, fertile **Paradise Valley,** a place of such beauty that some Hollywood stars have established homes here. One of the state's more popular hot springs is located near the town of *Prey,* Montana. The 100-year-old lodge at Chico Hot Springs is built like a farmhouse, with a grassy front lawn and a nearby stable. People flock here for two reasons: The huge pool is warm and soothing, and the resort has one of the finest kitchens in Montana. Dinner reservations are sometimes hard to come by in the summer.

Just before it tumbles into the broad Livingston plain, the Yellowstone River passes through a narrow canyon called Railroad Spur. Dam builders have cast covetous eyes on that gap for years. Any dam would flood one of the prettiest valleys in Montana. Fortunately, that plan is now dead. The Yellowstone remains the longest free-flowing river in America.

Laurel

We now return to Laurel, Montana, to follow the direct route to Livingston. The drive is not particularly interesting because you have to take the interstate, which follows along the north side of the river. Somewhere between here and Columbus—scholars are not quite sure of the exact spot—Clark's party paused for five days to make canoes. Private Gibson, who had received a severe wound from being bucked from his horse, found travel painful, so Clark decided that if they could find suitable trees, this would be a good place to take to the river.

July 19th, 1806 [Clark] I proceeded on about 9 miles, and halted to let the horses graze and let Gibson rest. his leg become So numed from remaining in one position, as to render it extreemly painfull to him. I derected Shields to keep through the thick timber and examine for a tree sufficiently large & sound to make a canoe, and also hunt for some Wild Ginger for a Poltice for Gibsons wound. he joind me at dinner with 2 fat Bucks but found neither tree or Ginger.

July 20th, 1806 [Clark] I directed Sergt. Pryor and Shields each of them good judges of timber to proceed on down the river and examine the bottoms of any larger trees. I also sent two men in serch of wood soutable for ax handles they found some choke cherry which is the best wood which can be precured in this country. Sergt Pryor and Shields returned and informed me that they had proceeded for about 12 miles without finding a tree better than those near my Camp. I deturmined to have two canoes made out of the largest of those trees and lash them together which will cause them to be Study and fully sufficient to take my small party & Self with what little baggage we have.

The canoe proved sturdy indeed; they didn't abandon it until three days out of St. Louis.

The Crows were not the only thieves in the area.

July 23rd, 1806 [Clark] last night the wolves came into our camp and eat the most of our dryed meat which was on a scaffold.

Columbus/Big Timber

A nice diversion from the tedium of the interstate is to drive into the little town of Columbus, where you'll find one of the more unusual saloons in Montana. Every wall in the place is adorned with an elk rack, moose head, or set of antelope horns, and display cases house some of the most unusual stuffed animals you have ever seen. The collection includes a stuffed "jackalope," but that's not unusual; you can see them in bars throughout the west. The back-bar is one of the finest you'll see in this state, which has a plethora of splendid back-bars. Fifteen miles southwest of Columbus is the delightful town of **Absarokee,** which the natives pronounce Ab-SOR-key. This is the gateway to the Absaroka wilderness area, one of the wildest sections of the entire Rocky Mountain chain. The hiking here is superb.

Big Timber, Montana, seems to have been named more for hope than fact. There are no forests in these parts, and even in 1806, when Captain Clark was desperate to find trees big enough to make canoes, all he could find were a few scrawny cottonwoods.

Greycliff Prairie Dog Town State Monument

Across the interstate from Big Timber is a fine example of one of the most unique cultures in the entire western plains. The Corps of Discovery came upon its first prairie dog town near Fort Randall, South Dakota. The men spent an entire day trying to capture one of the lovable little fellows.

> *Sept 7th, 1804 [Clark] Cap Lewis & Myself walked up to the top, discovered a Village of Small animals that burrow in the grown Called by the french Petite Chien. Cought one a live by poreing a great quantity of Water in his hole. the Village of those animals Covd about 4 acres of Ground and Contains great numbers of holes on the top of which those little animals Set erect make a Whistleing noise and whin allarmed Step into their hole. Those Animals are about the Size of a Small Squrel Shorter & thicker, the head much resembling a Squirel in every respect, except the ears which is Shorter, his tail like a ground squirel which they shake & whistle when allarmd, the toe nails long, they have fine fur & the longer hairs is gray.*

The captains no doubt kept the crew busy hauling water, for some prairie dog burrows are as much as 14 feet deep. The prairie dog town here, along the interstate, has about 30 "neighborhoods," each composed of six to eight adults and a couple of dozen children. Intrepretive signs point out a number of interesting things regarding these colonies. You learn a bit about what prairie dogs eat, who their predators are, how they communicate with each other, and how they are able to survive in a harsh climate with virtually no water. One of the more interesting aspects of prairie dog life at the time of Lewis and Clark was their symbiotic relationship with the buffalo. These members of the squirrel family require a long line of sight for protection from predators. The buffalo kept the grass short. If you take your time, and move cautiously, you can get within a few feet of an adult before his squeaking becomes more and more frantic, the bobbing of the tail more and more purposeful, and he finally dives for cover.

It was near here that the Corps of Discovery first suspected the existence of the Crows, though at this point they seem to have naively felt that once they were able to powwow with them, they would find them as friendly as the Shoshoni, Flatheads, and Nez Perce. As we saw earlier, that was not to be.

Livingston

For motorists traveling west on Interstate 90, Livingston is really the beginning of the resort part of Montana. The town was established by the railroad as the gateway to Yellowstone National Park, and it continued to serve that purpose well into the 1970s, when Amtrak unwisely opted for the northern route through Glacier Park. Only a short, 50-mile section of the northern line goes through pretty country during daylight hours. Here, on the old Northern Pacific route, rail passengers used to enjoy 350 miles of uninterrupted grandeur.

Livingston is a great town with lots of sports shops, cowboy stores, and good country-style restaurants. Numerous Wild West saloons grace the downtown streets, but that's what you would expect of the place where that straight-shootin', whip-crackin' frontierswoman, Calamity Jane, used to hang out. Seven thousand people live here now. Many work for the railroad or in the sawmill east of town. Every fly fisherman worth his salt knows about Dan Bailey's fly shop; it's probably the biggest tackle store in the country. Though Clark wasn't interested in sport fishing, he did comment: "Although just leaving a high snowy mountain, the Yellowstone is alread a bold, rapid, and deep stream, 120 yards in width."

Bozeman Pass

If there were any justice in this world, Bozeman Pass would be called Sacagawea Pass. Although she was not the pathfinder some zealots tried to make her out to be, she did in fact steer Captain Clark toward this pass, which separates the Yellowstone from the upper Missouri watersheds. The party had been working its way up the Gallatin River:

July 13th, 1806 [Clark] The country in the forks between Gallitins & Madisens rivers is a butifull leavel plain covered with low grass. I observe Several leading roads which appear to pass to a gap of the mountain in a E.N E. direction about 18 or 20 miles distant. The indian woman who has been of great service to me as a pilot through this country recommends a gap in the mountain more south which I shall cross.

The party crossed the divide without any difficulty whatsoever.

July 15th, 1806 [Biddle] After an early breakfast they pursued the buffalo-road over a low gap in the mountain and at the distance of six miles reached the top of the dividing ridge which separates the waters of the Missouri and the Yellowstone: It now appeared that communication between the two rivers was short and easy. From the head of the Missouri at its Three Forks to this place is a distance of 48 miles, the greatest part of which is through a level plain; with an excellent road over a high, dry country, with hills of inconsiderable height and no difficulty in passing.

John Bozeman was the next white man to take advantage of this "short and easy" communication. In 1863 he selected this pass for his wagon road to the Montana mines. The Bozeman road left the Oregon trail 100 miles northwest of Laramie, Wyoming, and paralleled present-day Interstate 25 to Billings, where it then followed Interstate 90 to the Three Forks. Jim Bridger never liked the idea of the Bozeman road. He thought the Sioux were still strong enough to keep the white man out, and he was right. John Bozeman got himself killed four years later for his trouble. His grave is nearby, along with an itinerant prospector named Henry T. P. Comstock, who found some gold near Virginia City, Nevada, but left town before anyone knew that the Comstock Lode would produce enough riches to finance the Union Army during the Civil War. In order to avoid the Crow and the Cheyenne, Jim Bridger thought it best to run a road west of the Big Horn and the Wind River Range. That route was ultimately chosen for the narrow-gauge Utah Northern Railroad. The Wyoming portion of the Bozeman road was closed in 1868 after a brigade of soldiers under the command of Brevet-Lieutenant Colonel William F. Fetterman marched into a trap set by Chief Red Cloud. Seventy-nine officers and soldiers died in that melee. Only six were killed by bullets; the rest were killed by arrows, hatchets, or spears and were tortured before death. Fetterman and his second in command were found together, evidently having shot each other to prevent capture and torture.

The Bozeman Pass won out in the end, however, when Fredrick Billings chose it for the Northern Pacific Railroad. He had a regiment of soldiers of the United States Army stationed at nearby Fort Ellis to keep the peace. The pass, of course, is not over the Continental Divide; it simply saves about a 200-mile detour north along the Missouri. The always windy summit is 5800 feet above sea level.

Bozeman

The city of Bozeman has always been a lively place. In the 1890s the town had an opera house that imported productions from New York City. It's the home of **Montana State University,** the second largest in the state. Many would argue that this is the prettiest college town in the United States. Twenty-two thousand people live here. Cattle ranching and wheat farming are big industries, but tourism is, too. The Gallatin River comes down out of Yellowstone and makes a swing to the west here. The headwaters of that river are near West Yellowstone, which, during the summer, is a lively place, because it has the airport closest to the park with commercial flights. Forty miles south of town is **Big**

Sky, a first-class resort developed by the late TV anchorman Chet Huntley. With numerous hotels and condominiums, Big Sky is Montana's poshest ski area. The **Jim Bridger Ski Resort** is 15 miles north of town.

The MSU campus is on the southern edge of town. To get there, take South Willston Avenue, which leads past ten grand old houses on the National Register of Historic Places. The campus boasts an enormous field house that seats 13,000 people for various sporting events. In a nearby building is the **Museum of the Rockies,** with its diverse collection of rare Indian and western art, antiques, and an authentic recreation of a western store and blacksmithy. The museum, however, is most noted for its collection of dinosaur fossils, which have recently been featured on the television science program "Nova."

Madison Buffalo Jump State Park

To continue west, you can either take the interstate or duck off on the old highway that goes from Bozeman to Three Forks. The latter is substantially more charming. From the town of *Logan,* a gravel road goes seven miles south to the Madison Buffalo Jump. An intrepretive exhibit demonstrates the hunting technique that Lewis and Clark saw in the Missouri Breaks, whereby prehistoric man was able to stampede buffalo herds over a cliff. At the town of Three Forks we once again pick up the westward trek of the Corps of Discovery.

THE CONTINENTAL DIVIDE

The elation the captains felt upon reaching the Three Forks was short-lived; they now had to decide which fork to follow. Clark's anxiety begins to show in his writing. He had expected to find the Shoshoni here, and although he apparently had a bad case of heat stroke, he went out looking for them anyway. He should have stayed in camp.

July 26th, 1805 [Clark] I deturmined to leave Shabono & one man who had Sore feet to rest & proceed on with the other two to the top of a mountain 12 miles distant west from thence view the river & vallies a head, we with great dificuelty & much fatigue reached the top at 11 oClock. We Contind thro a Deep Vallie without a Tree to Shade us scorching with heat. I was fatigued my feet with Several blisters & Stuck with prickley pears. I felt my Self verry unwell & took up Camp on the little river 3 miles above its mouth & near the place it falls into the bottom.

Clark walked 21 miles that day. When Lewis found his comrade in such straits he ordered a three-day halt and administered to his patient:

July 27th, 1805 [Lewis] I prevailed on him to take a doze of Rushes pills, which I have always found sovereign in such cases and to bath his feet in warm water and rest himself.

Lewis and Clark Caverns

The Corps of Discovery, after leaving the valley of the Three Forks, entered a rather narrow canyon.

August 1st, 1805 [Lewis] the mountains are extreemly bare of timber and our rout lay through the steep valleys exposed to the heat of the sun. I felt my sperits much revived on our near approach to the river at the sight of a herd of Elk of which Drewyer and myself killed two. we

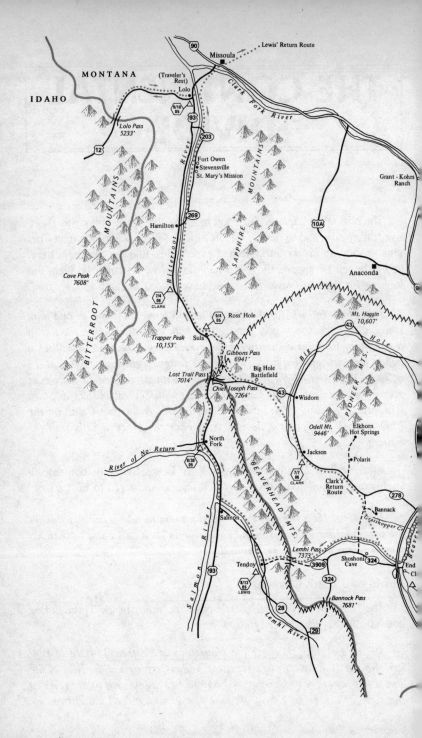

THE CONTINENTAL DIVIDE

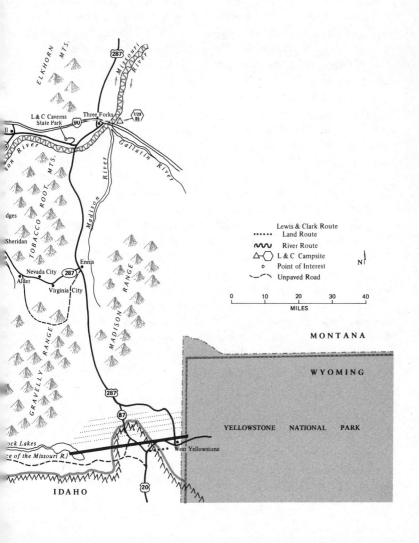

made a comfortable meal of the Elk and left the ballance of the meat on the bank of the river for the party with Capt Clark. this supply was no doubt very acceptable to them as they had had no fresh meat for near two days except one beaver Game being very scarce and shy.

They had no way of knowing it, but they were passing underneath what was later to be named **Lewis and Clark Caverns,** said to be the third largest in the country. Discovered by a local rancher, Don Morrison, in 1902, the caves are now administered by the Montana State Parks system. A paved but steep road leads three miles up the hillside to the park headquarters, where there is a lovely view of the river below, a cafe and gift store, and an intrepretive center. The ranger-guided walk through the cave takes about 2 hours, and is well worth the small fee. You see bats, stalactites, stalagmites, hidden pools, and a thousand grotesque shapes and figures as you wend your way, sometimes on hands and knees, through numerous intricate passageways.

The mountains on the south are the **Tobacco Roots,** those to the north are the **Elkhorns.** The valley below, fortunately, wasn't so narrow it could not be used by the railroads. The Northern Pacific follows the north bank, the now defunct Milwaukee, St. Paul and Pacific the south. The latter route was electrified, giving it a big fuel-saving advantage over its rival. When the bankers shut it down, one wonders why the Burlington Northern didn't snap it up. As you emerge from the canyon near the town of **Whitehall,** you enter the 75-mile-long **Beaverhead Valley,** one of Montana's most scenic. The crew of the Corps of Discovery that was paddling, polling, and pulling this tiny flotilla probably didn't notice the scenery or care. The weather had become hot, the water cold, the current swift, and so much brush crowded the banks they couldn't use a tow rope. Progress could only be made by having the boatmen wade in the water and push the canoes ahead of them. If this wasn't bad enough, the river meandered so much, the distance traveled by water was often two or three times that made by land.

August 2nd, 1805 [Lewis] this is the first time I ever dared to wade the river. The valley is from 6 to 8 miles wide and consists of a beautifull level plain with but little timber. the land is tolerably fertile, and is either black or dark yeallow loam, covered with grass from 9 inches to 2 feet high. the plain ascends gradually on either side of the river to the bases of two ranges of high mountains. the tops of these mountains are yet covered partially with snow, while we in the valley are nearly suffo-

cated with the intense heat of the mid-day sun; the nights are so cold that two blankets are not more than sufficient covering.

As you look around this countryside today, you realize that the land must be "tolerably fertile." Whitehall, the first town you come to in the Beaverhead Valley, is a quiet little village with a one-block main street and a 1940s-style motel that is still well maintained. A couple of miles farther on, at the little hamlet of **Waterloo,** Lewis and Clark lost their last chance to effect an easy crossing of the Continental Divide. The Minnetarees had suggested a route up present-day Pipestone Creek, which would have led them over a pass of that name. This was the approximate route chosen by the railroad 80 years later, because it offered an easy grade both up to the 6400-foot summit and then down to the headwaters of Clark Fork River, a tributary of the long-sought Columbia. Instead, they continued south, desperately looking for the Shoshoni, and in the process added 100 extra miles to their journey. The present-day follower of Lewis and Clark is well advised to make a detour here and visit a bit of country the explorers missed.

Butte City The richest hill on earth, Butte City has been called, made famous by mines with names like Neversweat, Leonard, Mountain Consolidated, Orphan Girl, Emma, Belmont, High Oregon, Tramway, Mountain View, and the richest of them all, the fabulous Anaconda. Marcus Daly, the Irish immigrant whom we met at Helena, came from Calaveras County, California, in 1880. He had been a foreman at the legendary Comstock in Virginia City, Nevada, and later held the same job at Alta, Utah, site of the present-day ski resort. Daly was uneducated, hard driving, and shrewd; he was a man of charm and wit and was well liked, but also had a fiery temper. Having sold his interest in a Butte silver mine for $100,000, he bought the Anaconda for $30,000. Its shaft was only 60 feet deep. By the time the miners had drilled 300 feet it became obvious that Daly had a copper mine, not a silver mine. With partners from his San Francisco days, James Ben Ali Haggin, George Hearst (William Randolph's father), and Lloyd Tevis, he had, by 1884, not only the best mine in the district, but also the world's largest smelter, 28 miles to the west in a town he named after his mine. Daly probably would have owned Montana were it not for a man quite his antithesis, William Clark.

Clark got here first. He was from Pennsylvania, had attended Laurel Hill Academy, and had spent two years at Ohio Wesleyan studying law. He arrived at the gold fields in Bannack, Montana, "a little red-headed man with a pack on his back." Like Leland Stanford before him, he quickly learned that merchandising was the road to fortune. In this business he asked for no quarter, and gave none, but he had no advantage save his intelligence and determination. Recognizing the value of the mines in Butte, he went east to the Columbia School of Mines, where he took a cram course in mineralogy. He came back, bought a stamp mill, and was well on the road to wealth when Daly arrived in 1876. In 1888 Clark ran for territorial delegate to Congress and won the Democratic primary, which in those days was tantamount to election. Inexplicably, Marcus Daly saw things differently; he paid a lot of people to suddenly become Republicans, and Clark lost the general election. The feud between them was to last the rest of their lives. Clark ultimately got to the Senate in 1900, but was forced to resign because he got the votes the same way Daly did. Clark's home in Butte, a Victorian-Elizabethan masterpiece built in 1888, is now called the **Copper King Mansion**. It is a privately run museum, open to the public for a fee. Marcus Daly's likeness can be seen on the campus of the **Montana School of Mines,** where a statue pays tribute to his help in founding the school.

Butte has a rather sad look to it today. Some time ago the mines turned to open-pit operation and the ever expanding **Berkeley Pit** threatened to engulf the downtown, but copper prices plunged and mining ceased in 1982. With a population of 37,000, Butte is the third largest city in Montana. Unless industry can be attracted here, however, it's liable to go the way of a thousand other boom-bust towns in the west. The city now has one of the highest unemployment rates in the country.

Butte City is doing all it can to promote tourism. A motorized replica of an old Butte streetcar makes a tour of the city and its environs during the summer. A stop is made at the viewing stand overlooking the Berkeley Pit, the largest open-pit copper mine in the world. Twenty billion pounds of copper were taken out of that hole. The bus makes a stop at the **World Museum of Mining,** a nice place with a rather extravagant name. The exhibits are housed in an old tin building next to the head-frame of the Orphan Girl mine. Numerous pieces of machinery are shown, including an old mine locomotive that is now used to take visitors around the site. Nearby, **Hell Roarin' Gulch** is a reconstructed old-west mining town, complete with barber shop, bank, pharmacy, Chinese laundry, and, of course, a saloon that, unfortunately, is not open for business.

A stroll through downtown Butte is enjoyable; there are more than 40 buildings of historical interest, and the streets above the hill are dotted with old mansions, one of which has been converted into a bed-and-breakfast establishment and another into an art gallery. A fine collection of minerals is housed at the college.

Anaconda, located 28 miles away, is also having some hard times. The downtown, which is 3 miles west of Interstate 90, has a number of fine old brick buildings, several of which have been converted into a new shopping mall. The 1898 **county courthouse,** located on the hillside south of downtown, is especially handsome. The **Copper Village Museum and Art Center** houses traveling art shows and has exhibits of early pioneer life. The public library was a gift of Phoebe Apperson Hearst, wife of the publisher William Randolph Hearst.

Deer Lodge The old stone-walled **State Prison,** now open to the public, is located in Deer Lodge, about 30 miles north of Anaconda. This town of 4500 also has an **antique car museum** said to house the largest collection of Ford automobiles in the world. For those who want to get a more genuine feel for the old west, go north a mile or so to **Grant-Kohrs Ranch**. The heirs of the last operators of this ranch decided to preserve some of the life and lore of a big-time, million-acre cattle spread. The family donated the main house, outbuildings, and 200 acres of land to the National Park Service, which conducts daily guided tours. The emphasis here is on describing what cowboy life was actually like. You see a real bunkhouse, a real blacksmith shop, and some things you don't find in Hollywood movies—an ice house, for example. A few farm animals are kept on the premises to enhance the romantic image presented here.

Robbers Roost/Nevada City/Virginia City

Returning to the voyage of Lewis and Clark, the first town you come to south of the Pipestone Creek is *Twin Bridges,* where the river again forks. The river coming in from the west is the Big Hole, which the captains called the Wisdom. We will explore a town named Wisdom and the upper reaches of the Big Hole river at the end of this chapter.

Entering from the east is a river Lewis and Clark called the Philanthropy. Although enough gold was removed from that river to endow a dozen sizable philanthropic foundations, the name didn't stick. It's now the Ruby River. One of its tributaries, Alder Creek (pronounced crick), was a forgotten place in 1863 when six miners, led by Bill Fairweather, did a little prospecting. One of the party, Henry Edgar, kept a journal:

> "Fairweather and I were to make camp and stand guard. The other four proceeded up the gulch . . . prospecting. About sundown Bill went

across the creek to picket the horses. 'There is a piece of bedrock projecting,' said Bill, 'and we had better go over and see if we cannot get enough money to buy a little tobacco.' While Edgar washed the first pan of dirt, Fairweather scratched around in the bedrock with his butcher knife and . . . called: 'I've found a scad.' I had the pan about half washed down, and I replied: 'If you have one I have a thousand,' and so I had.''

Within three months 10,000 people were swarming over Alder Gulch's dry, sagebrush-covered hills. By the end of the first year $10 million worth of gold was taken out of Alder Creek, making it one of the richest placer mines ever discovered.

To reach the gold fields, turn southeast at Twin Bridges on State Highway 287. The little town of **Sheridan,** population 650, has a stately feeling. A few miles farther on you come to the infamous roadhouse, Robber's Roost, sometime home of the notorious Henry Plummer and his band of "Innocents," so called because their password was "I am innocent." Henry Plummer was smooth and shrewd, yet he was a likable man, even though he had a criminal record covering some ten years in the west. He arrived at the Montana gold fields in the winter of 1862, and by spring he had pulled enough wool over enough eyes to win an election for sheriff. With the law on his side, he and his bloody gang of 24 murdered 102 known victims, most of them on their way to **Bannack City** to sell their "dust." Robber's Roost was aptly named. A sign in front of the building reads: "The main floor was a shrine to Bacchus and Lady Luck. The second floor was dedicated to Terpsichore and bullet holes in the logs attest to the fervor of ardent swains for fickle sirens."

By December 1833 the honest miners had had enough. They organized Montana's first "Miners Court," the so-called Vigilante Committee. Civil matters were determined by a jury, but in criminal cases whoever showed up got a vote, and that usually meant everyone in town. The problem was how to mete out justice; there were no prisons and few jails. Punishment, therefore, ranged from flogging (rare), banishment (common), and hanging (predominant). The first to suffer the miners' revenge were a trio: Paris Pfouts, James Williams, and Wilbur Sanders. By summer 45 had been hanged, 24 here at Robber's Roost. Historians don't know why, but a mysterious sign "3–7–77" was placed on all the corpses. The Innocents got their 3–7–77s when the Vigilantes caught one Erastus "Red" Yeager, who "fingered" his pals. Henry Plummer met his maker on January 10, 1864, in the nearby town of Bannack.

While sheriff, he himself had constructed the gallows they used. Robber's Roost is now an antiques shop.

A few miles up the river is Nevada City, once described as "a hell-roaring town where men were men and women were scarce." When the gravels gave out it became a ghost town, but now it is a thriving tourist center. In the mid-1940s the few remaining buildings were fixed up and store shelves were stocked with period merchandise. As interest in the project grew, old buildings from all over western Montana were moved here. The result is a beautiful re-creation of how people used to live, an American equivalent of Oslo's Bygdoy Park. Several buildings have been put to modern use—you can stay in the **Nevada City Hotel** and eat at the old-time bakery. One building houses a fantastic collection of coin-operated calliopes. For a small fee you can tour through a Chinese laundry, a grocery store, a superintendent's house, a firehouse, a school, and view what's billed as the world's only two-story "out house." Across the street is a restored train station, with a number of old passenger cars parked on the sidings. From here you can take a little narrow-gauge train to the neighboring mining camp, Virginia City.

Nevada City doesn't have a mayor, so it can't echo something that happened in its namesake city in California. It seems that Nevada City, California, was founded well before the neighboring territory became a state. The mayor wrote the governor of Nevada, complaining that too many people were getting the places confused, and, therefore, demanded that the state change its name to something else.

Virginia City was the territorial capital for a while, and later a county seat, so it never became a ghost town. In fact, one mine and a ball mill are still being worked in a halfhearted sort of way. The main street is full of tourist shops, but there are businesses catering to the locals, and the town hasn't lost its gold-camp feel. The town boasts two playhouses, one for legitimate theater, the other for melodrama and olio acts. One hotel has been built to look like a bunch of tumbledown miners' shacks, but inside the rooms are clean and modern. You can get a glass of "red-eye" at a saloon sporting a beautiful back-bar, or drink sarsaparilla at an "Olde" ice-cream parlor. The **Madison County Museum** houses a collection of turn-of-the-century artifacts and shows a bit of the history of the mining enterprises. North of town is "boot hill," resting place for five men—George Lane, Boone Helm, Frank Parrish, Haze Lyons, and Jack Gallagher, all of whom became "retired road agents" when they were hanged on January 14, 1864.

Gravelly Range Tour The Ruby River's headwaters are in the Gravelly Range, due south of Virginia City. The **Beaverhead National Forest** encourages visitors to take an all-day tour of these mountains, not only to see some beautiful scenery, but also to gain a better understanding of good forest and range-land practices. The Forest Service publishes a *Self-Guided Tour* brochure that describes what to see along the way. It takes a full day to drive this 60-mile gravel road, much of which is quite steep, but you truly get out into the backcountry where you'll see elk, moose, deer, antelope, and grouse, and learn how the sheepherder "tipis out," a practice whereby he moves his flock to a new pasture each day. On a clear day you can see Wyoming's far-off Grand Tetons from the 9500-foot summit of this lovely drive. The tour starts in *Alder* and terminates in *Ennis*. Inquire at the ranger station in either Sheridan or Ennis before making this loop.

Beaverhead Rock

At Twin Bridges the Jefferson River loses its identity and becomes the Beaverhead. You enter Beaverhead County, which includes Beaverhead National Forest and a prominent landmark called **Beaverhead Rock**. Sacagawea gave us these names.

August 8th, 1805 [Lewis] the Indian woman recognized the point of a high plain to our right which she informed us was not very distant from the summer retreat of her nation on a river beyond the mountains which runs to the west. this hill she says her nation calls the beaver's head from a conceived resemblance of it's figure to the head of that animal.

Beaverhead Rock, located about 10 miles south of Twin Bridges, is now a State Historical Monument. The best view is had by driving past the bluff and parking just beyond the point where the highway crosses the river. This, incidentally, is also a good place to look at the river and ponder what life must have been like for the Corps of Discovery. Imagine a party of 32, still burdened with a substantial amount of baggage, trying to row, pole, and pull their canoes up this icy little stream.

August 12th, 1805 [Lewis] this morning Capt. Clark set out early. found the river shoally, rapid, shallow, and extreemly difficult. the men in the water almost all day. they are geting weak soar and much fortiegued; they complained of the fortiegue to which the navigation sub-

jected them and wished to go by land. Capt. C. engouraged them and passifyed them.

It's little wonder that it took the main party three weeks to go the hundred miles from Three Forks to the place where they finally abandoned their boats and waited for the Shoshoni to bring them horses. Lewis, knowing that he had to increase the pace, set out on a plan.

August 8th, 1805 [Lewis] I determined to proceed tomorrow with a small party to the source of the principal stream of this river and pass the mountains to the Columbia; and down that river untill I found the Indians; in short it is my resolution to find them or some other, who have horses if it should cause me a trip of one month. for without horses we shall be obliged to leave a great part of our stores, of which, it appers to me that we have a stock already sufficiently small for the length of the voyage before us.

About halfway between Beaverhead Rock and Dillon you cross the 5000-foot contour, and you're almost a mile above sea level.

Dillon

Dillon is the center of trade for the whole southwest corner of the state, giving it a downtown that is much more active than its 4000 population could normally support. The captains recognized the agricultural potential of this valley:

July 10th, 1806 [Biddle] The valley is open and fertile; besides the innumerable quantities of beaver and otter with which its creeks are supplied, the bushes of the low grounds are a favorite resort for deer; while on the higher parts of the valley are seen scattered groups of antelopes, and still further, on the steep sides of the mountains, are observed many bighorns, which take refuge there from the wolves and bears.

The courthouse spire dominates the Dillon skyline. Nearby, Beaverhead County maintains a lovely museum next to the railroad tracks. The large collection of branding irons is especially interesting. Not surprisingly, the beautiful old sandstone railway depot is no longer used for its intended purpose. The railroad through here is the Union Pacific, but it wasn't always so. It began life in 1863 as a narrow-gauge road called the Utah Northern, built by John W. Young, Brigham Young's son. The route then, as now, connected the great mines of Bannack, Butte, and Virginia City with the mainline at Ogden, Utah. It was widened to standard gauge in 1888 and later became part of Henry Villard's Oregon Short Line, which in turn was absorbed into the Union Pacific. An even prettier building is at **Western Montana College,** a brick-turreted masterpiece that has somehow survived the wrecker's ball. The headquarters of Beaverhead National Forest is here, so if you are planning to follow Lewis and Clark over the Lemhi Pass, it's a good idea to check in to determine the condition of the road.

Clark Canyon Dam

Just south of Dillon you rejoin Interstate 15. After about 20 miles of driving turn right on the county road, which crosses Clark Canyon Dam. Proceed up the hill about a mile to a spot called **Camp Fortunate Overlook.** An event occurred here that even a master dramatist would have trouble conceiving. Recall that Lewis had gone on ahead to look for the Shoshoni. He had now returned with a band of that nation, including its chief, a young man named Cameahwait. Here, the explorers tell the story of what happened.

August 17th, 1805 [Biddle] they had not gone more than a mile before Clark saw Sacajawea, who was with her husband 100 yards ahead, begin to dance and show every mark of the most extravagant joy, turning around him and pointing to several Indians, whom he now saw advancing on horseback, sucking her fingers at the same time to indicate that they were of her native tribe. a woman made her way through the crowd towards Sacajawea, and recognising each other, they embraced with the most tender affection. Captain Clark went on, and was received by Captain Lewis and the chief, who after the first embraces and salutations were over, conducted him to a sort of circular tent or shade of willows. After this the conference was to be opened. Glad of an opportunity of being able to converse more intelligibly, Sacajawea was sent for; she came into the tent, sat down, and was beginning to interpret, when in the person of Cameahwait she recognised her brother. She instantly jumped up, and ran and embraced him, throwing over him her blanket and weeping profusely.

It turns out that Lewis had had a hard time getting Cameahwait to accompany him back to this spot. The tribe was desperately fearful of the better-armed Blackfeet, and the chief thought he was being led into a trap. The presence of his sister among the group put that thought to rest. The following day was Lewis's birthday, and he seems to have had one of his bouts with melancholy.

> *August 18th, 1805 [Lewis] This day I completed my thirty first year, and conceived that I had in all human probability now existed about half the period which I am to remain in this Sublunary world. I reflected that I had as yet done but little, very little, indeed, to further the hapiness of the human race or to advance the information of the succeeding generation. I viewed with regret the many hours I have spent in indolence, and now soarly feel the want of that information which those hours would have given me had they been judiciously expended. but since they are past and cannot be recalled, I dash from me the gloomy thought, and resolved in future, to redouble my exertions and at least indeavour to promote those two primary objects of human existence, by giving them the aid of that portion of talents which nature and fortune have bestoed on me; or in future, to live for* mankind, *as I have heretoofore lived for* myself.

Lemhi Pass

To continue following Lewis and Clark's trail, you must backtrack in time to early August, when Captain Lewis had gone ahead to look for the Shoshoni. He realized that the time had finally come for the Corps of Discovery to give up the river.

> *August 10th, 1805 [Lewis] we arrived in a handsome open and leavel vally where the river divided itself nearly into two equal branches; here I halted and examined those streams and readily discovered from their size that it would be vain to attempt navigation of either any further.*

The stream junction Lewis was referring to is now beneath the water impounded by the Clark Canyon Dam. We, too, now leave the river and drive west, over the little-used Lemhi Pass. Following Lewis and Clark's trek here means really getting off the beaten path and into the backcountry. It's a wonderful experience, for two reasons: the scenery is spectacular, and you get a real sense of adventure. The road is paved for the first 20 miles, but then turns to gravel for another 20. In good weather there is no problem, even for today's low ground-clearance automobiles, but nevertheless it is wise to check with the Forest Service

Captain Lewis Meeting the Shoshone: Charles M. Russell. Russell chose the dramatic moment when Sacagawea made her tearful reunion with the tribe she had been abducted from at the age of sixteen. The warrior on horseback may be her brother, Cameahwait. Courtesy of the Thomas Gilcrease Institute of American History and Art, Tulsa, OK.

in Dillon, Montana, or Salmon, Idaho, before attempting a crossing. There is some marshy country to traverse on the Montana side, which could mean a deeply rutted road after a heavy rain. The road, of course, is closed in winter, and it may be well into spring before the snow melts. Be aware also that the road up the Idaho side is very steep.

If you decide against traveling the Lemhi Pass route, you have two choices. The paved county road you are following turns south to cross **Bannock Pass.** Though sections of this road are also gravel, the distance is shorter, and the grade is easier. The other choice is to forget following Lewis and Clark's outward journey across the Continental Divide and instead retrace the route taken by Captain Clark on his homeward trip. We'll explore that country at the end of this chapter.

Lewis, with three men—Drouillard, Shields, and McNeal—set out, cautiously working their way up Prairie Creek, keeping an especial eye out for signs of the Shoshoni. County Road 324 roughly parallels their walk. Somewhere in this large meadow, which they called Shoshoni Cove, perhaps at a point 10 miles from Interstate 15, they encountered the first person they had seen since leaving North Dakota. Lewis tried to show friendliness toward this Shoshoni warrior, but when Shields and Drouillard, not realizing his presence, kept advancing, he became suspicious, whipped his horse, and fled. The explorers were on foot and could not keep up. Nightfall found them about 10 miles east of the pass, near a formation called Red Butte. They were now more than 6000 feet above sea level, a fact that Lewis seems to have arrived at empirically.

August 10th, 1805 [Lewis] the mountains do not appear very high in any direction tho' the tops of some of them are partially covered with snow. this convinces me that we have ascended to a great hight since we have entered the rocky Mountains, yet the ascent has been so gradual along the vallies that it was scarcely perceptable by land.

The next day the tiny party set out early. To follow in their footsteps, continue on the county road about 3 miles past the point where it swings toward the south. A small sign identifies the road to Lemhi Pass, which is Forest Service Route 3909. The road first passes through a marsh teeming with migratory birds, and then wends its way through grazing land and past an isolated farmhouse. Wildlife seem to be everywhere—squirrels, rabbits, deer, blackbirds, and grouse. You soon enter a canyon and begin to climb, although not at an alarming rate. After passing another ranch you turn a corner and there, up ahead, you can see what must be the pass. A sign points to a spot that has become a bit of a shrine for Lewis and Clark buffs.

> *August 12th, 1805 [Lewis] at the distance of 4 miles further the road took us to the most distant fountain of the waters of the Mighty Missouri in surch of which we have spent so many toilsome days and wristless nights. thus far I had accomplished one of those great objects on which my mind has been unalterably fixed for many years, judge then of the pleasure I felt in allaying my thirst with this pure and ice-cold water which issues from the base of a low mountain or hill. two miles below McNeal had exultingly stood with a foot on each side of this little rivulet and thanked his god that he had lived to bestride the mighty & heretofore deemed endless Missouri.*

The spring still runs, and you can straddle the creek if you like. Unfortunately, it is not "the most distant fountain" of the Missouri. That honor goes to a small stream that feeds into Upper Red Rock Lake, which, incidentally, is also a nice place to visit. **Red Rock Lakes National Wildlife Refuge** was established primarily to protect the rare trumpeter swan, largest of all North American waterfowl. To reach the refuge, drive 40 miles south from Clark Canyon Dam on the interstate, and then turn east at the town of **Monida** on a dirt road that is open from mid-April to mid-November. It is a 28-mile drive to the refuge headquarters.

Above the spring the road to Lemhi Pass begins to climb steeply, but in a half mile you're at the summit. This is the exact spot where Meriwether Lewis stepped foot outside of the United States, thereby becoming the first American to cross the Continental Divide. It's beautiful. The mountains seem to go on endlessly toward the west, presenting what Lewis must have considered to be a formidable barrier. Except for a split-rail fence that marks the Idaho border, the place looks almost exactly as it did when the explorers stood here, though you do see one of those comforting Lewis and Clark Trail signs that you have been following since leaving Illinois. Here, for the first time, it points the way to a westward flowing river. You are at 7373 feet, the highest elevation Lewis and Clark would see, and you are smack atop the Continental Divide. In spite of the beauty and serenity here, there's a bit of sadness, too. The explorers deserved better luck. They had certainly earned it, but it turns out they could not have picked a worse spot to cross the Rockies. Lewis is about to descend into country we now call "The River of No Return."

August 12th, 1805 [Lewis] I now descended the mountain which I found much steeper than on the opposite side, to a handsome bold running Creek of cold Clear water. here I first tasted the water of the great Columbia river. after a short halt we continued our march along the Indian road to a spring where we found sufficient quantity of dry willow brush for fuel, here we encamped for the night having traveled about 20 Miles. as we had killed nothing during the day we now boiled and eat the remainder of our pork, having yet a little flour and parched meal.

On the Idaho side the dirt road is quite steep for the first couple of miles, but then it joins a branch of the Lemhi River, which it follows to its junction with the main river near the one-building hamlet of Tendoy, Idaho. You're back in civilization now; the valley is wide and fertile and dotted with farmhouses. It's easy to understand why the Shoshoni chose this spot for their home because it's cozy, with high mountains on three sides and an impassable river on the fourth. You could feel safe here, though, strangely, there was no game. The Shoshoni lived on a meager diet of berries, dried bulbs, and pounded salmon, and only occasionally crossed the Lemhi Pass to hunt the buffalo.

Tendoy

Lewis's party finally met face to face with the Shoshoni on the 13th. They had surprised three elderly women, who thought they were Blackfeet. By the time the warriors who had been sent to rescue them arrived, Lewis had succeeded in allaying the women's fears by offering them some trinkets. The braves, whose cosmetic tastes included smearing their faces with bear grease, greeted the explorers warmly.

August 13th, 1805 [Lewis] these men advanced and embraced me very affectionately in their way which is by puting their left arm over your wright sholder clasping your back, while they apply their left cheek to yours. bothe parties now advanced and we wer all carresed and besmeared with their grease and paint till I was heartily tired of the national hug. all the women and children of the camp were shortly collected about the lodge to indulge themselves with looking at us, we being the first white persons they had ever seen.

Fortunately, the Shoshoni had horses. "Many of them would make a figure on the South side of the James River," Lewis commented. They had mules, too—more prized than horses by these mountain people because of their sure-footedness on steep trails. Both had been obtained from other Shoshoni tribes to the south, who in turn had obtained them from the Spanish. Although this band had never seen white men, they were aware of their existence. Lewis was fascinated by this isolated tribe. His journal goes to great lengths describing their dress, social mores, eating and hunting habits, and their relationships with strangers. Being the consummate scientist, he was anxious to learn if they suffered from venereal disease, which he considered a white man's malady. The answer was yes, but only to a small degree, leaving him to suspect that it might have been transmitted from the whites by a neighboring tribe.

Lewis had to use every bit of diplomacy at his command to convince the Shoshoni to return to the Beaverhead with him, so that they could meet up with the rest of the party. He was successful, as we have seen, and Cameahwait had his tearful reunion with his sister. After the reunion Lewis sent Clark on ahead to reconnoiter the Lemhi and Salmon rivers, while he busied himself cacheing the canoes and building saddles for the newly purchased horses. By August 29 the entire party was over the divide and encamped on the Lemhi River.

Twenty years later fur trappers would move into the Lemhi River country, but they seemed to have had no better luck than those at Three Forks. In 1823 the Hudson's Bay Company's Finnan MacDonald and a party of 51 men were camped a few miles east of the present highway when they were ambushed by a band of Blackfeet. MacDonald lost five men before they were able to regroup and set fire to the thicket where the Blackfeet were hiding. Sixty-eight warriors died in the inferno. MacDonald was later heard to say: "The beaver will have a golden skin before I trap this valley again."

The Mormons, who tried to establish a colony here in 1855, named the river after King Limhi in the Book of Mormon. Brigham Young dispatched a party of 27 here with instructions to teach the Indians "the arts of husbandry and peace according to our Gospel plan." They had no better luck than the Catholics and the Presbyterians did with other tribes. The Shoshoni wanted the white man's magic, but not his religion. A relief party was sent out from Salt Lake City in April of 1858, and the fort the Mormons had built was abandoned. A stone monument, located on a bluff overlooking the valley, commemorates these events. The monument is 50 yards off a country road, a half mile east of the main highway, about 2 miles north of Tendoy. Sacagawea was born here, and there once was a monument commemorating that event, but it's gone now, vandalized by an uncaring public, so the locals say.

Salmon

Considering the ruggedness of this country, the valley floor between Tendoy and Salmon, Idaho, is surprisingly gentle. Alfalfa fields and meadows grace the lowlands, while the nearby hills are covered with sagebrush. There is a hint, though, that the vegetation is different from what you saw on the east side, because you pass the first of the many sawmills you'll see on your way to the Pacific. The Lemhi River loses its name 20 miles south of Tendoy, where you enter the town of Salmon. This town of 3500 has a well-deserved reputation as a sportsman's paradise. It has a real western feel, with real western saloons. Beer is the preferred drink these days, not red-eye. Pickup trucks come from all over the county to cruise Main Street after dark. No one would be seen without a deer rifle hung from a rack by the rear window. Life here is not all macho, though. Salmon has 13 churches, two theaters, a bowling alley, and a nine-hole golf course. Lemhi County operates a nice little historical museum in the center of town.

The River of No Return

Clark went exploring the lower river with a Shoshoni guide named Toby, while Lewis remained in the village, learning all he could from the chief. The two captains, each in his own way, came to the same conclusion: There was no way they could continue to follow this river.

August 13th, 1805 [Lewis] Cameahwait informed me that this stream discharged itself into another doubly as large but he added on further enquiry that there was but little timber below the junction and that the river was confined between inaccessable mountains, was very rapid and rocky, insomuch that it was impossible for us to pass either by land or water down this river to the great lake where the white men lived as he had been informed. this was unwelcome information but I still hoped that this account had been exaggerated with a view to detain us among them.

August 20th, 1805 [Lewis] the Chief further informed me that he had understood from the persed nosed Indians who inhabit this river below the rocky mountains that it ran a great way toward the seting sun and finally lost itself in a great lake of water which was illy taisted, and where the white man lived.

> *August 23rd, 1805 [Clark] proceed on with great dificuelty as the rocks were So sharp large and unsettled and the hill sides Steep that the horses could with the greatest risque and dificulty get on. I proceeded on. Sometimes in a Small wolf parth & at other times Climing over the rocks for 12 miles the river is almost one continued rapid, five verry considerable rapids the passage of either with Canoes is entirely impossible. my guide and maney other Indians tell me that the Mountains Close and is a perpendicular Clift on each Side, and Continues for a great distance and that the water runs with great violence from one rock to the other on each Side foaming & roreing thro rocks in every direction, So as to render the passage of any thing impossible.*

To this day no one has ever tried to build a road down through that abyss. A dirt road goes west from the town of **North Fork** to the confluence of the Main and Middle Salmon rivers; that's all. It's 153 miles from North Fork to the next paved road at **Riggins.** You can now get there by river, though, thanks to modern technology, which was spurred on by people's lust for thrills. It's for good reason that the town of Salmon bills itself as the "White Water Capital of the World."

Both the North and Middle Forks have been declared "Wild and Scenic Rivers," so no motorboats are allowed. Numerous outfitters offer trips ranging in duration from an afternoon to several weeks. Normally, you take your personal belongings, tent, and sleeping bag. The outfitter provides the food and cooks the meals. Parties are usually limited to about 15. A great way to see most of the Salmon wilderness area is to take one of the trips that floats both the Middle and Main Forks, and takes about 11 days. You are met at Salmon and flown into the put-in point at **Indian Springs.** It's a "white knuckle" flight in a 6-passenger, single-engine plane, but your pilot is likely to have two things going for him: He's reasonably old and he spent most of his life flying "bush" in Alaska. You don't get to be old in that business unless you are a very good pilot. The Middle Fork is especially interesting for fly fishermen because it's a "catch and release" stream, teeming with cutthroat trout. President Jimmy Carter took a much-photographed vacation shooting these rapids. The Main Salmon sees migrating steelhead trout and chinook salmon. All along the way there are idle times of peaceful floating, coupled with thrilling wet times where your heart is in your mouth, and you emerge from the torrent with a tongue feeling like cotton. The take-out point is **Riggins,** Idaho, where you're flown back to Salmon.

Lewis and Clark found themselves thwarted, and the alternatives didn't seem much better. Lewis's indomitable spirit, however, seems to have come forth.

August 20th, 1805 [Lewis] I now asked Cameahwait by what rout the Pierced nosed indians came over to the Missouri; this he informed me was to the north, but added that the road was a very bad one as he had been informed by them and that they had suffered excessively with hunger on the rout being obliged to subsist for many days on berries alone as there was no game in that part of the mountains which were broken rockey and so thickly covered with timber that they could scarcely pass. I felt perfectly satisfyed, that if the Indians could pass these mountains with their women and Children, that we could also pass them; and that if the nations on this river below the mountains were as numerous as they were stated to be that they must have some means of subsistence which it would be equally in our power to procure.

Lost Trail Pass

So, on September 1, with winter fast approaching, the party started trudging up the North Fork of the Salmon, heading back toward the Continental Divide. Here they got their first taste of what real western mountains are like. The highway up the North Fork of the Salmon and over Lost Trail Pass is wide and fast, but it soon becomes obvious that the Corps of Discovery was going to have a time of it. This isn't a gentle, U-shaped valley like the one they found at the Lemhi; the canyon walls here are steep. Clark's journal has an angry tone to it.

September 2nd, 1805 [Clark] proceded on thro' thickets in which we were obliged to Cut a road, over rockey hill Sides where our horses were in perpeteal danger of Slipping to their certain distruction & up & Down Steep hills, where Several horses fell, Some turned over, and others Sliped down Steep hill Sides, one horse Crippeled & 2 gave out. with the greatest dificuelty risque &c. we made five miles & Encamped.

September 3rd, 1805 [Clark] This day we passed over emence hils and Some of the worst roads that ever horses passed, our horses frequently fell Snow about 2 inches deep when it began to rain which termonated in a Sleet.

The next morning they reached the Lost Trail Pass. They had been in Idaho 19 days, during which time they had descended 2700 feet, trekked along 60 miles of river, climbed back up 2300 feet, and were no closer to their objective than if they had simply made a two-day tramp west from Twin Bridges. But they did have horses.

Sula

The north side of Lost Trail Pass is not quite as steep as the south and the party was able to negotiate it in one day. They didn't have to recross the Continental Divide here, though it's only a mile away and not much higher in altitude. Near present-day Sula, Montana, in a valley called **Ross's Hole,** they quite unexpectedly met a band of Flatheads.

September 4th, 1805 [Clark] prosued our Course down the Creek to the forks about 5 miles where we met a party of the Tushepau nation, of 33 lodges about 80 men 400 Total and at least 500 horses, those people receved us friendly, threw white robes over our Sholders & Smoked in the pipes of peace, I was the first white men who ever wer on the waters of this river.

So the Shoshoni weren't the only friendly tribe that had horses. Lewis referred to this tribe as the Tushapaws, but in the same paragraph also called them flat heads. A common misconception about this name should be cleared up at this point. Lewis and Clark were to meet a small number of natives who used boards to flatten their heads, but those tribes were on the Pacific Coast, not here in western Montana. These people got their name from the hand motions that translators used to define the tribe when conversing in sign language. They spoke with a gurgling, throaty sound, in a language the captains had never heard before. Conversation was difficult, but they found a Shoshoni boy living among them who translated to Sacagawea. She rendered the words into Minnetaree for Charbonneau. He, in turn, translated them into French for Labiche, who passed them on in English to Lewis and Clark. Because of the new language and the tribe's lighter than normal skin, several of the soldiers thought they had finally found the lost tribe of Welshmen, which John Evans had sought among the Mandans in 1796–1797. They were, however, of a Salishan linguistic stock common in the Pacific Northwest, and had lived here for centuries.

You're at the very headwaters of the Bitterroot River. Captain Lewis was indirectly responsible for the name.

> *August 22nd, 1805 [Lewis] another speceis [of root] was much mutilated but appeared to be fibrous; the parts were brittle, hard, of the size of a small quill, cilindric and as white as snow throughout. this the Indians with me informed were always boiled for use. I made the experiment, found that they became perfectly soft by boiling, but had a very bitter taste, which was naucious to my pallate, and I transfered them to the Indians who had eat them heartily.*

Lewis received credit for his discovery; the scientific name for the tuber is *Lewisia rediviva*. The upper part of the Bitterroot River runs through **Ross's Hole,** which was named after Alexander Ross, a Hudson's Bay Company trapper who, with his wife and child, got caught in a snowstorm here in 1824 and nearly perished. He called this the "Valley of Trouble." It is about as beautiful as anyone could imagine. Charlie Russell captured some of the magic of the place in his celebrated painting adorning the House of Representatives Chamber at the Capitol in Helena. The scene depicts the Corps of Discovery on its return trip. The broad valley is now planted to alfalfa and pasture, with a few farmhouses sprinkled about. Here and there you see what seems to be the symbol of western Montana's charm, the sloped ramp of a hay lifter. These 60-foot-high structures, built of lodgepole pine logs that slant upward at a 50° angle, are used to pile hay into great mounds. There is a ranger station and a grocery store/gas station in Sula, but little else. People come here to fish the river, hunt in the fall, and, in general, just get away from it all.

Downstream from Ross's Hole the river enters a narrow canyon where you come upon a stand of ponderosa pine. This behemoth of a tree is common to the eastern, or drier slopes of the western mountains. The ponderosa, together with its cousins the Jeffrey, white, and yellow pines, make up the backbone of the timber harvest in this area. After a few miles of threading your way through the canyon, you come out onto the 75-mile-long Bitterroot Valley. It has much of the same character as the Lehmi Valley, and to a lesser extent the Beaverhead, but here the mountains are nearer, and they look gigantic. By Colorado standards, they're not high, but some of them soar over a mile above this 4000-foot valley. A sign along the roadway points out **Trapper Peak,** at 10,153 feet the highest in the Bitterroots. The little town of **Darby,** population 500, has two sawmills, one of which makes "Lincoln Log" type homes.

As you motor along this pleasant road, it's easy to forget that even though the Corps of Discovery now had horses, the going wasn't all that easy. Today, you cross a lateral stream on a concrete bridge or over an iron culvert without even noticing it, but before the highway engineer worked his magic, even an insignificant brook could be a major obstacle.

July 4th, 1806 [Biddle] We therefore went on till at the distance of a mile we came to a very large creek, which, like all those in the valley, has an immense rapidity of descent; we therefore proceeded up for some distance, in order to select the most convenient spot for fording. Even there, however, such was the violence of the current that, though the water was not higher than the bellies of the horses, the resistance made in passing, caused the stream to rise over their backs and loads.

Hamilton

Only 2600 people live in Hamilton, but it is rapidly gaining a reputation as a great place to live. The Chamber of Commerce boasts that it has the best climate in western Montana because the Bitterroot Mountains block off the big storms coming in from the Pacific. Even Meriwether Lewis commented that it was warmer here than along the same latitudes near the Atlantic, but since his last thermometer had been smashed coming over the Lost Trail Pass, he couldn't prove it. The fact that there are two bookshops, four sporting goods stores, and a French restaurant tells a bit about the people who live in these parts. Old brick buildings are being rehabilitated; upscale stores are moving in. It's always been a classy place. Our friend Marcus Daly had a ranch nearby where he bred the only Montana horse ever to win the Kentucky Derby. Unfortunately, the home is not open to the public.

The old **courthouse,** nestled among some lovely residences on tree-lined streets, is now a fine museum. There is an extensive collection of Indian artifacts, many of which depict the life and lore of the Flathead nation. One room houses an exhibit showing how scientists struggled to find the cause and a cure for Rocky Mountain Spotted Fever. This typhuslike disease, carried by ticks, is endemic to the area. Vaccinations now control its harmful effects.

Instead of taking the main highway north from Hamilton cross the railroad tracks and follow County Road 373. On his return trip, Captain Clark liked what he saw along here.

July 3rd, 1806 [Clark] after letting our horses graze a Sufficient length of time to fill themselves, and taking dinner of venison we again re-

sumed our journey up the Vally which we found more boutifully versified with small open plains covered with a great variety of Sweet cented plants, flowers & grass.

July 4th, 1806 [Clark] This being the day of the decleration of Independence of the United States I had every disposition to selebrate and therefore halted early and partook of a Sumptious Dinner of a fat Saddle of Venison and Mush of Cows [roots].

This lovely rural area used to be extensively planted with apple orchards, but, according to the curator at the museum in Hamilton, the ranchers found they couldn't compete with Washington State's Yakima Valley. The latter was on the well-run Great Northern Railway, while the products grown here had to be shipped east on the Northern Pacific. The product was often spoiled before it got to market.

Stevensville

The little town of Stevensville is often called Montana's cradle because it is the site of the first permanent settlement in the state. The Nez Perce had sent no fewer than four delegations to St. Louis in 1831, 1835, 1837, and 1839, begging for the whites to come and bring their "Black Robes," meaning their medicine. The Protestants came in 1836, but didn't stay. Finally, in 1841, the Jesuits sent Father De Smet and two other priests, Nicholas Point and Gregory Mengarini, who were guided here by William Ashley's mountain man, Thomas Fitzpatrick. The three men established St. Mary's Mission, where they planted the first agricultural crops in the state—oats, wheat, and potatoes. In 1845 they built a gristmill out of stones quarried in Belgium and shipped to Fort Vancouver, from whence they were brought overland by oxcart. The Nez Perce and Flatheads did not see in the cross a symbol of the crucifixion, with its attendant corollary meanings, but rather conceived of it as a symbol of physical power. Father Anthony Ravalli took over the mission in 1845, but became a bit discouraged when, as he put it, the natives resorted "to savage obscenity and shameless excesses of the flesh." De Smet went off to convert the Blackfeet, which enraged the Flatheads. What good was medicine if your enemies had it, too? In 1850 Father Ravalli sold the mission to Major John Owen for $250, but then returned in 1866 and established the present-day St. Mary's Mission. Thanks largely to his skills as an artist, pharmacist, and carpenter, he won the respect of the tribe, who worshiped him until his death in 1884. The county is now named in his honor. You can tour the church he built and visit his home, both of which are administered by the Missoula Diocese of the Catholic Church.

Meanwhile, Major Owen was conducting a thriving fur trade from the stockade he called Fort Owen. He married a Flathead named Nancy, but when she died demon rum got the best of him, and he was deported to Philadelphia at the request of the governor. Remnants of the old fort have been preserved and now house exhibits and mementos of the period. The buildings are located about a half mile northwest of Stevensville.

Meriwether Lewis tried to learn as much about the country from the Flatheads as he could.

September 9th, 1805 [Lewis] our guide [the Shoshoni, Toby] could not inform us where this river discharged itself into the columbia river, he informed us that it continues it's course along the mountains to the N. as far as he knew it and that not very distant from where we then were it formed a junction with a stream nearly as large as itself which took it's rise in the mountains near the Missouri to the East of us and passed through an extensive valley generally open prarie which forms an excellent pass to the Missouri. the point of the Missouri where this Indian pass intersects it, is about 30 miles above the gates of the rocky Mountains *the guide informed us that a man might pass to the missouri from hence by that rout in four days.*

Lewis now knew the awful truth. What had taken the Corps of Discovery 52 days to traverse, they could have done in four. His decision to go home via this route must have been an easy one to make.

Lolo (Travelers Rest)

The Corps of Discovery paused to rest at the mouth of a little stream coming out of the mountains from the west. They named this place Travelers Rest. As you stand here today looking out over the valley, it's a pleasant sight, but upon turning around and facing west, what do you see? Condominiums. This pristine river junction is rapidly being swallowed up by Missoula, 10 miles to the north.

Travelers Rest is a good place to pause even today, because the trip over the Bitterroots is a long one, and there is almost no place to stay between here and Lewiston, Idaho, 210 miles away. It is also time to go back to the Beaverhead Valley and pick up Captain Clark's homeward journey.

The Big Hole Country

As noted earlier, an alternate to crossing the Lemhi Pass is to follow the route taken by Captain Clark on his return in 1806. We pick up this trail at Clark Canyon Dam, where Lewis had cached the canoes the previous fall. The men of the Corps of Discovery had a special reason to hurry back to this spot.

July 8th, 1806 [Clark] the most of the Party with me being Chewers of Tobacco become so impatient to be chewing it that they scercely gave themselves time to take their saddles off their horses before they were off to the deposit. The country through which we passed to day was diversified high dry and uneaven Stoney open plains and low bottoms very boggy with high mountains on the tops and North sides of which there was Snow,

One wonders whether the men would have been in such a hurry if they had known that the previous night they had camped directly atop one of the biggest gold fields ever discovered. Lewis and Clark were superb explorers and scientists. They saw and identified numerous species of animals, birds, plants, flowers, and trees. They even spotted a bit of the mineral wealth of the upper Missouri, the coal seams near Fort Peck and along the Powder River. If they had only scratched the surface of the gravels in Grasshopper Creek, they would have returned to St. Louis rich men indeed. Grasshopper Creek was the site of Bannack City, first capital of Montana Territory. You can drive there in one of two ways. The route most closely following Clark's return is to drive 9 miles west on the road to Lemhi Pass, but then turn north on a 12-mile dirt road that leads to Bannack State Park. Perhaps a better route is to return toward Dillon and then take paved State Highway 278 over a gentle pass that drops down into Grasshopper Creek. A well-graded gravel road leads 4 miles south to the State Park.

Bannack

Bannack is about as fascinating a ghost town as you'll see anywhere in the west because, just as at California's Bodie, there is not a hint of commercialism anywhere. The dry desert air has prevented the buildings from rotting, so though weeds grow in the main street and sagebrush in the front yards, the place otherwise looks exactly as it did in the mid-1860s. Gold was discovered on July 28, 1862 by John White, William Eads, and a small party of prospectors from Colorado. By the following winter 500 miners were poking among the gravels, and by spring there were 1000. They named the town for the Bannock Indians, but didn't get the spelling quite right. Two years later Henry Plummer returned from Virginia City to make his date with the vigilantes and the history books. The scaffold he had built and was forced to use still stands in a little valley a quarter of a mile north of town. Plummer's last words were a plea for "a good drop."

The existence of the vigilantes focused attention on the need for law and order, which in turn required some sort of territorial status. The miners sent one Sidney Edgerton to Washington to plead their case. Whether the gold Edgerton took with him had any bearing on the outcome, we do not know. What we do know is that Congress acted, and President Lincoln named him the first governor of Montana Territory. Edgerton's home, now called the **Governor's Mansion,** is open to visitors. The first courthouse, built of brick, was later turned into a hotel, which continued to accept guests until the 1940s. Peeling wallpaper now hangs from cracked plaster walls. All told, about 60 structures still survive, including the Masonic Temple, jail, Methodist church, saloon, and a cabin used by bootleggers only 20-odd years ago.

Bannack went into a near panic the night of August 12, 1877, when the beating of drums signaled the presence of Chief Joseph and his band of Nez Perce, fresh from their "victory" at the Big Hole. Women and children were sheltered in the hotel, where extra food, water, and bedding were assembled. General Howard was summoned, and earth and log breastworks were constructed, but of course the only thing the Nez Perce wanted was to escape from the white man's clutches. From there the hearty band retreated into present-day Yellowstone Park, and thence to the Bear's Paw Mountains, where General Howard caught up with them two months later.

Elkhorn Hot Springs

Clark's return trail almost exactly follows today's County Road 278, which skirts the southern extremity of the Pioneer Mountains. A reasonably challenging ski resort has been built on Maverick Mountain near **Polaris,** Montana, a town that comprises a total of two buildings: a school and a post office. The road to Polaris is gravel, but well graded. Nearby, you can stay at one of the state's numerous hot-spring resorts. The old timbered building at Elkhorn has all the charm of a 1930s-type resort, with overstuffed chairs surrounding a giant fireplace. The swimming is relaxing in the 100° water.

Wisdom

The county road crosses a gentle pass that Captain Clark learned about from Sacagawea.

July 6th, 1806 [Clark] we assended a small rise and beheld an open beutifull Leavel Vally or plain about 15 Miles wide and near 30 long extending N & S. in every direction around which I could see high points of Mountains covered with snow. The Squar pointed to the gap through which she said we must pass.

The beautiful plain he is referring to is the valley of the Big Hole River. This is about as typical a range land as you'll find anywhere in Montana. Don't be surprised if you have to pull off the road and park while a couple of ranch hands drive a herd of whiteface from one range to another. Two tiny towns dot the valley, **Jackson** and Wisdom. Wisdom, which carries on the name that Lewis and Clark gave to the Big Hole River, is an interesting place to stay because you get a real feel for the Montana outback. Ranch hands whoop it up in the local bar, and the highlight of the restaurant's menu is Chicken Fried Steak. A local fellow, an entrepreneur or outlaw depending on your point of view, raises coyotes for their pelts. The animals howl most of the night. Some of the best fishing in the Rockies is along the Big Hole, so the only store in town carries a huge supply of flies and lures.

Big Hole National Battlefield

Ten miles west of Wisdom, nestled beneath Battle Mountain, is the place where Captain Clark's party came out of the mountains.

July 6th, 1806 [Biddle] In the afternoon we passed along the hillside north of the creek, till, in the course of six miles we entered an extensive level plain. Sacajawea recognized the plain immediately. She had traveled it often during her childhood, and informed us that it was the great resort of the Shoshonees, who came for the purpose of gathering quamash and cows, and of taking beaver, with which the plain abounded.

Seventy-one years later the Nez Perce followed the same route. They, too, felt they had found sanctuary in this "great resort of the Shoshonees." Having escaped from General Howard's army in Idaho, the band, under the leadership of Chief Looking Glass, paused to rest and recapture their spirit. They chose a spot where the steep slopes of the mountain merged with the willow-dotted marsh that stretches to the river bed. The southeast bank of the Big Hole River is grassy here. For the first time since leaving the Clearwater River, they could set up their tipis. They cut trees to replace lost or broken poles, and by the afternoon of August 9 had erected 89 lodges. When the work was done they played games, sang songs, danced, and told stories long into the night. Looking Glass, believing that they were at last out of danger, did not post guards. What he didn't know was that Colonel Gibbon and the 7th U.S. Infantry was hard on their trail.

Gibbon planned a dawn ambush, which got a premature start when a marksman shot a lone Nez Perce who was going after his horse. The soldiers waded across the river and began firing indiscriminately into the tipis, killing about 40 women and children before the warriors had even located their guns. The Nez Perce soon found sniping positions, though, and forced the soldiers back across the river and into a forest, where they were pinned down by deadly accurate fire. The siege continued throughout the day and night and into the next day. By the afternoon of the 10th many warriors began withdrawing to help Chief Joseph gather the tribe, care for the injured and dead, and round up the horses. By nightfall they fired some parting shots and left with the remaining women and children. General Howard's lumbering army arrived an hour later and found Gibbon's command decimated, with 29 dead and 40 wounded. In a military sense, the Nez Perce had won the battle, but the win was hollow. They lost about 30 warriors, staggering losses for such a small band, and even more shattering was the final realization that the army was not going to leave them in peace.

The site of the battle is now administered by the National Park Service. An intrepretive center tells much of how the battle was fought, but the most moving experience is to walk among the trees and look out at the grass where the teepees were located, imagining what must have occurred on those two horrid days in August.

Gibbons Pass/Chief Joseph Pass

As you continue west, the highway begins to climb, and soon you're in a pine forest. To follow Clark's route as closely as possible, 9 miles west of the battlefield turn north on a gravel Forest Service road that leads to Gibbons Pass. The road skirts some of the loveliest meadows you have ever seen. The little meandering brook abounds with rainbow trout. After driving about 7 miles you come to the place where Captain Clark returned to the United States.

July 6th, 1806 [Clark] Some frost this morning the last night was so cold that I could not sleep. we set out and proceeded up the creek 3 Miles and left the road which we came on last fall to our right and assended a ridge with a gentle slope to the dividing mountain which Seperates the waters of the Middle fork of Clarks river from those of Wisdom and Lewis's river,

The rise up to the summit has been so gentle that it is hard to believe that you're astride the Continental Divide. Two days later Captain Clark wrote of this route:

July 8th, 1806 [Clark] The road which we have traveled from travellers rest Creek to this place is an excellent road. This road and with only a few trees being cut out of the way would be an excellent waggon road. one Mountain of about 4 miles over excepted which would require a little digging. The distance is 164 miles.

A hundred years later, Lewis and Clark's chronicler, Elliott Coues would muse:

[Coues] It seems almost incredible that the modesty or the indifference of the great explorer should have led him to dismiss this part of his route without further remark. A road for 164 miles, fit for wagons except at one point, across the great Continental Divide—we hardly realize what it meant to make that discovery in 1806.

Gibbons Pass, of course, is named for the fellow who massacred all those women and children at the Big Hole. The road down the other side is a bit steeper than on the east side, but as you drive it you can see that Captain Clark was right about it being an easy route. The dirt road rejoins the Corps of Discovery's 1805 route at the town of Shula.

From there to Travelers Rest, the outbound and homeward bound routes cover identical ground. If you are uneasy about driving this backcountry Forest Service Road, stay on the paved highway that crosses into the Columbia watershed at Chief Joseph Pass, which is just a mile east of the Lost Trail Pass, the one Lewis and Clark crossed on their westward journey. The elevation here is 7141 feet. From the summit take U.S. Highway 93 to Travelers Rest.

THE BLACKFEET COUNTRY

When Thomas Jefferson bought Louisiana the understanding was that it included all the land drained by the Missouri River. It was incumbent upon Lewis and Clark, therefore, to find out as much as they could about where that boundary might lie. On the outbound trip in 1805 the explorers unexpectedly came upon a river that they named Marias, after Lewis's cousin in Virginia. They hoped that the source river would prove to be far to the north because then much of what they suspected was rich fur country would belong to the United States, not England. On their return trip in 1806 Meriwether Lewis decided to have a look around. The party had split at Travelers Rest (Missoula), with Clark going south to reclaim the canoes and explore the Yellowstone River while Lewis took a shortcut to the Great Falls. From there he headed north to the Marias. We'll follow Lewis's route in this chapter. Keep in mind that they are coming home, so the journal dates read in reverse order.

Two Medicine Country

We pick up the route in Havre, which you may recall is on the Milk River. Instead of taking U.S. Highway 87 south, continue west on the Hi Line, U.S. Highway 2. This is the route followed by the Great Northern Railroad. The area feels more Canadian than American. The strongest radio stations come out of Regina and Lethbridge, and you're more or less paralleling the Canadian Pacific Railway which, 100 miles north of here, wends it way through towns with wonderful names: Moose Jaw, Swift Current, and Medicine Hat. The country seems awfully flat and a bit monotonous. The only break in the skyline is the **Sweet Grass Hills,** an isolated group of mountains almost on the Canadian border. This is dry-farming country. To reduce wind-blown soil erosion, the crops are strip-planted; 60-foot-wide swaths are left fallow every other year, giving the country the appearance of a bold, regimental-striped tie.

Almost without knowing it, you pass out of the Milk River drainage

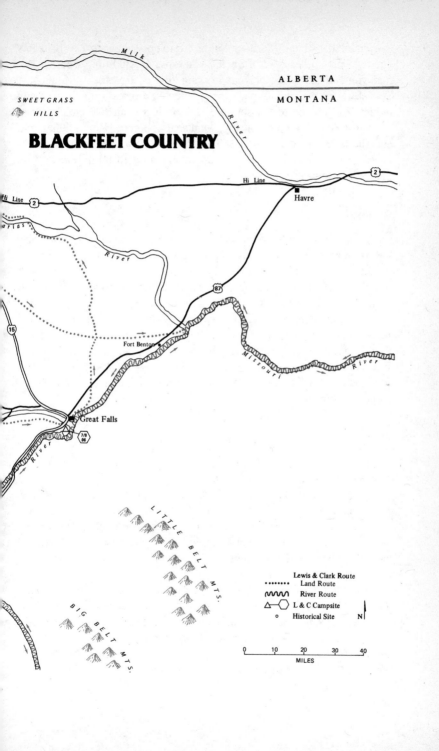

and enter that of the Marias. You're climbing, though it hardly seems like it. Thirty miles west of Shelby you cross the 3000-foot contour and find yourself straining for your first sighting of the Rocky Mountains. The sky and the clouds play tricks on your senses—are those really mountains you see to the southwest, or is it just an odd cloud formation? Just before reaching Shelby, Montana, you come atop a little knoll, and suddenly there is no doubt. The entire horizon looks like the working edge of a cross-cut saw. The mountains are still 60 miles away, so they don't dominate the scene yet.

Shelby

Shelby is a robust town that has gambled on land, oil, and prizefighting ever since the 1890s, when the Great Northern Railroad put it on the map. Though the population is only 3200, the downtown has a more lively feel to it than Havre. When the world goes one way Shelby seems to go the other. Around the turn of the century the local people, in need of some entertainment, hijacked a train carrying an opera troupe to Seattle. In the 1920s oil was discovered, and the town took on a betting mood. Jack Dempsey and Tommy Gibbons held their heavyweight title fight here on the Fourth of July, 1923. Today, the town is at the junction of U.S. Highway 2 and Interstate 15, which makes it a natural stopover point for many travelers.

Cut Bank

Thirty miles west of Shelby, near the town of Cut Bank, the Maria's River loses its identity; one branch becomes the Two Medicine River, the other the Cut Bank Creek. Lewis and the Fields brothers, Reuben and Joseph, and the Great George Drouillard arrived here the third week in July. Lewis had planned to bring six men, but the theft of seven horses by Indians near the Great Falls caused him to change his mind. Alert to possible hostile activities, he brought six horses for the four men.

July 21st, 22nd, 1806 [Lewis] I am induced to beleive that there are falls in these rivers somewhere about their junction. being convinced that this stream came from the mountains I determined to pursue it as it will lead me to the most northern point to which the waters of Maria's river extend which I now fear will not be as far north as I wished and expected. we continued up the river for 17 miles when we halted to graize our horses and eat; there being no wood we were compelled to make our fire with the buffaloe dung which I found answered the purpose very well.

Cut Bank is a lively town on the edge of the Blackfeet Indian Reservation. The Blackfeet are one of the richest tribes in the country, thanks to the leasing of oil and gas rights. In spite of their early reputation of being warlike, they became the first tribe to adopt a constitution for tribal self-government. Money received from the sale of land to form Glacier National Park was used to buy horses and today much of the land has been put under irrigation. With a population of nearly 4000, Cut Bank serves as the major center of commerce for the tribe. Two events of interest to Lewis and Clark followers happened nearby. One occurred at a place Lewis called Camp Disappointment, the other at a place locally known as the "fight site." Both are a bit hard to find, are on private land, and are on sometimes impassable roads. Local inquiry is advised before trying to visit either place.

Camp Disappointment

To reach Camp Disappointment, drive west on U.S. Highway 2 for about 20 miles and turn right on County Road 483. Two and a half miles north of the highway a farm road goes off to the left, fords a little creek, and heads out into some wheat fields. At a salt flat you jog north for a bit and then climb to the top of a small knoll. Here the road gives out completely, and you find yourself bouncing along over the prairie. From the top of the knoll you can see the exact same grove of cottonwoods that Lewis wrote about. Beyond, the skyline is dominated by the jagged, snow-capped peaks of Glacier National Park, lying 20 miles to the west.

July 22nd, 1806 [Lewis] we found no timber untill we had traveled 12 miles further when we arrived at a clump of large cottonwood trees in a beautifull and extensive bottom of the river about 10 miles below the foot of the rocky Mountains. I now have lost all hope of the waters of this river ever extending to N. Latitude 50°.

The next day he established a camp in the cottonwoods, where he stayed three nights. He was desperately anxious to get a latitude reading, but the weather didn't cooperate, and the men were getting a bit anxious about meeting the dreaded Blackfeet. Lewis sent the hunters out looking for game, which they found were "spooked." These wilderness-wise men knew someone was in the area.

July 25th, 1806 [Lewis] here they [his hunters] found some winter camps of the natives and a great number of others of a more recent date; we consider ourselves extreemly fortunate in not having met with these people.

July 26th, 1806 [Lewis] The morning was cloudy and continued to rain as usual tho' the cloud seemed somewhat thiner I therefore posponed [leaving] but finding the contrary result I had the horses cought and we set out biding a lasting adieu to this place which I now call camp disappointment.

We now know that the camp was established at 48°40′. For years the United States and England argued about where the border should be. The dispute led in part to the declaration of war in 1812. The Treaty of Ghent, signed in 1818, established the boundary across the plains at the 49th parallel, but it was not until the 1840s, during which time the slogan "54°40′ or Fight" dominated the political rhetoric, that the same boundary was pushed on to the Pacific.

You can get nearly the same sense of where Lewis camped without having to go off into the backcountry, by continuing west on Highway 2 until you find a monument built a century ago by the Great Northern Railroad to honor the explorers. The road here is absolutely spectacular. You're driving through what looks like dead-level wheat fields, yet you know you must be climbing. Near the hamlet of **Meriwether,** where several grain elevators seem to pierce the Montana sky, you cross the 4000-foot contour. The mountains ahead totally dominate the scene; it looks like you're driving toward one giant wall. Up ahead, you're sure the road must suddenly come to a stop, for there is no possible way to get through that rampart. There is, of course. Highway 2 skirts the southern boundary of Glacier National Park.

The Lewis and Clark monument marks the northernmost extremity of the expedition. To continue following their route you must return to Cut Bank. Before you do so, consider spending a few days visiting the Blackfeet Nation and Glacier National Park.

Browning, population 12,000, is the last prairie town on Highway 2. The highlight of a visit here is a tour through the **Museum of the Plains Indian & Craft Center,** located just west of town. This facility, operated by the Department of the Interior, Indian Arts and Crafts Board, has a permanent exhibition gallery that presents the richness and diversity of historic arts created by such tribes as the Blackfeet, Crow, Northern Cheyenne, Sioux, Assiniboine, Arapaho, Shoshoni, Nez Perce, Flathead, Chippewa, and Cree. Of special interest is a multimedia presentation illustrating the migrations of the various tribes over the centuries. An extensive gift shop features locally made crafts. Nearby is the privately run **Museum of Montana Wildlife,** which has dioramas that focus on Rocky Mountain animals and features an extensive collection of for-sale items by Montana sculptor Bob Schriver.

Glacier Park Browning is 16 miles from East Glacier. There, almost ka-slamb, the plains cease and the mountains begin. This is perhaps the most impressive mountain escarpment in the entire Rocky Mountain chain. The prairie rises to about 5000 feet above sea level. Less than 2 miles away peaks soar to 8000 feet, and a few miles farther west to 10,000. In 1914 the Great Northern Railroad chose this site for **Glacier Park Lodge,** a building that is as handsome as Ycllowstone's Old Faithful Inn or the Grand Canyon's El Tovar Lodge. The four-story lobby is supported by 60 Douglas fir logs, each more than three feet in diameter. The Blackfeet, who watched the construction, called the building "Oom-Coo-La-Mush-Taw," meaning "Big Tree Lodge." This lovely resort hotel, with its swimming pool, riding stable, and nine-hole golf course, is only open from mid-June to mid-September. Run by the Glacier Park, Inc., concession, it is staffed by college students from all over the country. Three other beautiful hotels and several motels are operated by the same company. Take the time to spend at least one night each at **Lake McDonald Lodge, Many Glacier Lodge,** and the majestic **Prince of Wales Hotel,** located just across the border in Canada's Waterton Park. Park rangers give interpretive talks and present movies or slide shows nightly. Both Many Glacier Lodge and Glacier Park Lodge have nightly entertainment performed by the staff. Two mountain chalets are operated European-style; you can only get there by walking or on horseback. Blankets and full meals are provided.

Grizzlies Glacier Park has what many consider to be the finest hiking trails in the Rockies, trails that offer you the best opportunity to see the greatest of all North American animals in its natural habitat. It seems as if grizzlies are all anyone talks about around here. The gift shops sell bells that you tie to your leg or ankle to keep a ringing sound going when you're off in the wilds. It's supposed to warn the animal of your presence, but many, with tongue in cheek, call them the bears' "dinner bell." Park rangers give advice on what to do if you encounter one, and they often close trails where bears have been sighted. Lewis and Clark quickly learned to respect these beasts, and you will too. Here are some excerpts from their journals dealing with the subject:

April 29th, 1805 [Lewis] about 8 A.M. we fell in with two yellow [white] bear; both of which we wounded; one of them made his escape, the other after my firing on him pursued me seventy or eighty yards, but fortunately had been so badly wounded that he was unable to pursue so closely as to prevent my charging my gun; we again repeated our fire and killed him.

May 5th, 1805 [Clark] I went out with one man Geo Drewyer & Killed the bear, which was verry large and turrible looking animal, which we found verry hard to kill we Shot ten Balls into him before we killed him, & 5 of those Balls through his Lights [lungs] This animal is the largest of the carnivorous kind I ever saw.

May 14th, 1805 [Lewis] In the evening the men discovered a large brown [grizzly] bear lying in the open. six of them went out to attack him, all good hunters, two of them reserved their fires as had been previously conscerted, the four others fired nearly at the same time and put each his bullet through him, two of the balls passed through the bulk of both lobes of his lungs, in an instant this monster ran at them with open mouth, the two who had reserved their fires discharged their pieces at him as he came towards them, boath of them struck him, one only slightly and the other fortunately broke his shoulder, this however only retarded his motion for a moment only, the men unable to reload their guns took to flight, the bear pursued and had very nearly overtaken them before they reached the river; the others reloaded their pieces, each discharged his piece at him as they had an opportunity. they struck him several times again but the guns served only to direct the bear to them, in this manner he pursued two of them seperately so close that they were obliged to throw aside ther guns and pouches and throw themselves in to the river altho' the bank was nearly twenty feet perpendicular; one of those who still remained on shore shot him through the head and finally killed him; they then took him ashore and butchered him when they found eight balls had passed through him in different directions.

August 2nd, 1806 [Clark] this morning a Bear of the large vicious species being on a Sand bar raised himself up on his hind feet and looked at us as we passed down near the middle of the river. he plunged into the water and swam towards us, either from a disposition to attack't or from the cent of the meat which was in the canoes. we Shot him with three balls and he returned to Shore badly wounded.

Is it any wonder the scientists named this beast *Ursus horribilis*? There is a lot to see and do in Glacier National Park for those who don't want to brave the bear-infested outback. This is one park where you can have a splendid time without an automobile. Amtrak's Empire Builder stops at both East and West Glacier. It's a pleasant stroll through a lovely garden from the East Glacier train station to the hotel. Porters bring your baggage over on wheelbarrows. Vintage 1930s White Motor Car buses meet the train at West Glacier and take you to Lake McDonald Lodge. These old 20-passenger limousines have tops that roll back so you can take in the sun and the scenery. The buses connect all the hotels and go to such beauty spots as Two Medicine Lakes and the spectacular "Going to the Sun Highway," which crosses the Continental Divide at 6600-foot **Logan Pass.**

Cruise boats ply the waters of Lake McDonald, Many Glacier Lake, Two Medicine Lake, St. Mary Lake, and Waterton Lake. Many people take an early boat to the far end of these lakes, walk along some lovely trails that lead into the backcountry, and then catch the afternoon boat home. Canadian customs agents check the passengers on Waterton Lake, since the only stop is in the United States.

Marias Pass The railroad crosses the Continental Divide at Marias Pass, which, surprisingly, at 5236 feet, is the lowest in the United States. You hardly climb at all from Glacier Park Lodge before plunging into the Flathead River valley. It's a bit out of the way, but if you're bent on seeing the most lovable looking of the local animals, the Rocky Mountain Goat, drive west on Highway 2 to a place called the "Salt Lick," located about 3 miles east of the village of Essex. In the spring and early summer dozens of goats come down out of the mountains to replenish their salt supply on an outcropping just above the river. The highway department has constructed an overlook where you can watch these beasts perform unbelievable acrobatic acts on a nearly vertical hillside. Incidentally, there is a lovely old hotel in Essex named after the author of *The Compleate Angler,* Izaak Walton.

Fight Site

We now return to Lewis's journey. Recall that there were two important sites near Cut Bank—Camp Disappointment and the Fight Site.

Shortly after the four men left Camp Disappointment Lewis's worst fears were realized. They ran into a party of eight warriors. Historians disagree over whether they were Blackfeet or Gros Ventres. Lewis knew he couldn't make a run for it, so he boldly marched up and tried to make peace. He presented them with some beads and medals that seemed to satisfy them, but the braves insisted on staying with them that night. During the evening Lewis carried on an extensive conversation, with Drouillard acting as intrepreter. They claimed to be part of a large band, accompanied by a white man, that was encamped at the foot of the Rockies. Another sizable band, they said, was hunting buffalo to the west and was expected to be at the mouth of the Marias in a few days. Here, Meriwether Lewis recounts what happened the following morning:

July 27th, 1806 [Lewis] This morning at daylight the indians got up and crouded around the fire, J. Fields who was on post had carelessly laid his gun down behind him near where his brother was sleeping, one of the indians sliped behind him and took his gun and that of his brother. at the same instant two others advanced and seized the guns of Drewyer and myself, J. Fields seeing this turned about to look for his gun and saw the fellow just running off with her and his brother's. he called to his brother who instantly jumped up and pursued the indian with him whom they overtook. R. Fields as he seized his gun stabed the indian to the heart with his knife the fellow ran about 15 steps and fell dead; of this I did not know untill afterwards. Drewyer who was awake saw the indian take hold of his gun and instantly jumped up and seized her and rested her from him but the indian still retained his pouch. his jumping up and crying damn you let go my gun awakened me, I reached to seize my gun but found her gone, I then drew a pistol from my holster and terning myself about saw the indian making off with my gun I ran at him with my pistol and bid him lay down my gun which he was in the act of doing when the Fieldses returned and drew up their guns to shoot him which I forbid. Drewyer having about this time recovered his gun and pouch asked me if he might not kill the fellow which I also forbid. as soon as they found us all in possession of our arms they ran and indeavored to drive off all the horses. I pursued them so closely that they could not take twelve of their own horses but continued to drive one of mine with some others: being nearly out of breath I could pursue no further, I called to them as I had done several times before that I would shoot them if they did not give me my horse and raised my gun, one of them jumped behind a rock and spoke to the other who turned around and stoped. I shot him through the belly, he fell to his knees and on his wright elbow from which position he partly raised himself up and fired at me, and turning himself about crawled in behind a rock. he over shot me, being bearheaded I felt the wind of his bullet very distinctly.

So Lewis had two fatalities on his hands. It was the only time on the entire voyage when the explorers had to resort to guns, and the incident would affect Blackfeet/white man relationships for decades. Lewis's immediate reaction was to round up as many horses as he could, both their own and the Indians', and hightail it out of there. The four men rode 63 miles before they felt far enough away to rest and let their horses feed. Two hours later they were off again, covering another 17 miles before nightfall. The pace slowed after dark, but continued until two in the morning. By that time they had ridden almost 110 miles. "We now turned out our horses and laid ourselves down to rest in the plain very much fatigued as may be readily conceived," Lewis wrote in his journal. The next morning they joyfully joined up with the Gass-Ordway boat party, which was enroute downriver from the Great Falls.

The Fight Site is even more difficult to find than Camp Disappointment. Go back toward Cut Bank and turn south on County Road 358. After about 15 miles you will cross Two Medicine Creek about 4 miles downstream from where they camped that fateful night. About a mile past the bridge the highway takes a slight jog to the west. Here a dirt road heads out through some irrigated land where, after about 3 miles, it ends atop a bluff overlooking the riverbottom. A plaque marks the spot where the fight took place, a short distance from the end of the road. This road is impassable, except for four-wheel drive vehicles during irrigation season.

Bynum

If you choose not to visit the actual site of Lewis's fight, drive south on U.S. Highway 89 instead of returning to Cut Bank. It's a beautiful drive along what most historians call the Old North Trail, but Montanans call it the Whoop-up Trail. It's the road early settlers used to import whiskey from Canada to sell to the Indians. The mountains are constantly in view to the west. Just south of Cut Bank Creek a highway marker describes a bit about Lewis's fight and makes a big point of the theory that the warriors were not Blackfeet, but Gros Ventres. This is fine farm country, with many of the fields irrigated. The hamlet of Bynum is perhaps typical of the area. Bynum has three businesses. A grocery store and post office occupy one side of the street; the old bank building on the other side is now a saloon. The railroad that passed through town has recently been torn up. "It put a big dent in our social life," a local muses. "Everyone used to come by on Tuesday afternoon to watch the freight train go by." One of California's better small wineries is named Bynum. An inquiry of the barmaid asking if she stocks that brand brings a comment from the fellow at the next stool: "I didn't know you served wine in this place." She extracts a bottle of red wine from the refrigerator to prove she does.

Great Falls

Highway 89 is the shortest route between Yellowstone and Glacier Park, so during the summer every other vehicle you see is either a camper, an RV, or a car towing a house-trailer. The little town of **Choteau** is a pretty spot with a little museum and a spanking-clean "auto court"-type motel. Highway 87 heads east here, more or less following Lewis's route along the Sun River to Great Falls. The party had come down out of the mountains in the Dearborn River watershed, but crossed north to the Sun because Lewis thought they would find more game. He was right.

July 8th, 1806 [Lewis] Jos. Fields saw two buffaloe below us some distance which are the first that we have been seen. saw some barking squirils much rejoiced at finding ourselves in the plains of the Missouri which abound with game.

July 9th, 1806 [Lewis] Joseph feilds killed a very fat buffaloe bull and we halted to dine

July 11th, 1806 [Lewis] the morning was fair and the plains looked beatifull the grass much improved by the late rain. proceeded with the party across the plain to the white bear Islands (Great Falls) my course through a level beautifull and extensive high plain covered with immence hirds of buffaloe. it is now the season at which the buffaloe begin to coppelate and the bulls keep a tremendious roaring. I sincerely beleif that there were not less than 10 thousand buffaloe within a circle of that place.

Like all the plains Indians, the Blackfeet depended on the buffalo not only for food, but for moccasins, leggings, robes, and tipis. The beast was room and board on the hoof. Local legend said that the buffalo came from a great hole in the ground. When the seemingly impossible happened—herds that once had numbered perhaps 50 million were hunted to near extinction—many Blackfeet believed that the white man had found the hole, driven the buffalo back in, and plugged it up. The legend tells a better story than what actually happened. Lewis and Clark used the buffalo the same way the tribes did, but later the beasts were taken only for their hide. In time even these became a drag on the market. The Indians were not innocent of fostering waste, but at the height of the slaughter in the early 1880s buffalo were being killed solely because the white man liked the flavor of their tongues. Carcasses were left to rot in the prairie sun. By 1900 only 20 wild bison were known to exist.

Rogers Pass

Lewis left most of his party at Great Falls to help with the portage of the canoes, which were being brought down from the Three Forks area by the group under the command of Sergeant Ordway. He and his three companions then headed north toward the Marias and his date with the Blackfeet. We now pick up his trail as he came east, dropping down from the Continental Divide. Montana Highway 200 goes from Great Falls to Missoula, more or less following the famous **Mullan Road,** the first wagon route over the northern Rockies.

John Mullan, a lieutenant in the United States Army, came west in 1853 with General Isaac I. Stevens. Stevens continued on to Washington, where he later became the territorial governor, but he left Mullan and 13 men near present-day Missoula to survey a route for a wagon road over the Flathead and Bitterroot mountains. The road was to connect the heads of navigation of two rivers: Fort Benton on the Missouri and Wallula on the Columbia. The project was delayed for several years, but in 1859 Congress appropriated $100,000 for its construction. Mullan put 150 men to the task; they completed it in a year. It was neither the first nor the last cost overrun Congress would face; the final bill came to $230,000. The total length of the "road" was 624 miles, though in many places there were so many washouts the bull-whackers and mule-skinners who had to drive their teams along it no doubt called it other things.

Highway 200 wends its way through rolling country and crosses a gentle divide separating the Sun and Dearborn rivers. The mountains begin to loom up ahead, but nowhere do they come close to matching the spectacular ramparts at East Glacier. For the most part these are gentle, round-topped mountains, covered with lodgepole pine. The road enters a V-shaped canyon and you suddenly begin climbing. In a few miles you're at Rogers Pass on the Continental Divide. The elevation is 5609 feet. The coldest official temperature ever recorded in the continental United States was measured at a mining camp near here. On January 20, 1954, the mercury plunged to $-74°$.

Lewis and Clark Pass

The Nez Perce, who had steered Lewis into this country, advised against taking this pass, perhaps because they considered the eastside canyon too steep. They suggested that the party take a different road, which crosses into the Missouri waters about 6 miles north of Rogers Pass. That gap is now called Lewis and Clark Pass, a strange name considering the fact that Clark never laid eyes on this country.

You have to walk a bit, and it takes about 3 hours to visit Lewis and Clark Pass, but the countryside is so pretty it's worth the time and effort. About 7 miles west of Rogers Pass you come to Highway 279, which goes to Helena. A mile beyond that junction is Forest Service road number 293, which goes up Hardscrabble Creek. The road is unpaved and the going is slow in spots, but even a modern sedan will have no trouble negotiating the occasional rut. You cross into a lovely meadow, pass some isolated ranch houses, and begin to work your way up Alice Creek. The road ends 10½ miles from the main highway. Be careful—there is a fork 100 yards from the end. Don't take the more obvious route to the left, but continue straight ahead toward a clump of lodgepoles. Here you'll find a locked gate where you park your car. The 2-mile walk to the pass is really nothing more than a stroll. Wild flowers and ground squirrels are everywhere as you wander up along sloping fields of grass. The road steepens for the last quarter of a mile, but then you find yourself at 6421 feet above sea level on the wide level grassy plain where Meriwether Lewis and his party stepped back into the United States.

July 7th, 1806 [Lewis] Set out at 7 A.M. with the road through a level beatifull plain. much timber in the bottoms hills also timbered with pitch pine. no long leafed pine since we left the praries of the knobs. continued up it on the left hand side halted to dine at a large beaver dam. the hunters killed 3 deer and a fawn. passing the dividing ridge between the waters of the Columbia and Missouri rivers at ¼ of a mile. from this gap which is low and an easy ascent.

A sign marks the spot where they supped that day. The view west from Lewis and Clark Pass is lovely, but to get a good look at the prairies to the east you'll be well-advised to climb a bit more. **Red Mountain,** at 7100 feet, is to the north, and the 7450-foot **Green Mountain** lies a mile south. Add an extra hour or so to your schedule for either excursion. The Forest Service Ranger Station near Lincoln, Montana, sells a detailed map of this area.

Lincoln

The countryside west of Rogers Pass alternates between lovely meadows and stands of western white pine. These are the tribal lands of the Flatheads, but the river you begin to follow is named for the Blackfeet because they used this route for war parties against the Flatheads and Nez Perce. This rather low-key resort country is where people from Great Falls and Missoula have their summer homes. Exactly halfway between those two cities, you come to **7-Up Pete Creek,** location of an amazing restaurant. Here, nestled in a forest, practically out in the middle of nowhere, is a supper club and an adjoining dining room capable of seating 180 people at one time. Montanans, it seems, don't mind driving a few miles for supper if the food is good. A country-music band plays in the lounge most nights of the week during the summer.

The town of Lincoln, population 500, lies in an area Lewis called the Prairie of the Knobs. This pleasant resort community has several motels that cater to hunters and fishermen who want to get away from the mainstream tourist places. Forest Service roads fan out in all directions, providing easy access to numerous fishing spots and hunting areas. The elevation here is about 4500 feet, so the summer weather is ideal.

Hellgate

Highway 200 meets Interstate 90 at the confluence of the Blackfeet and the Clark Fork rivers. The nearby town of ***Milltown*** is—you guessed it—a mill town. This is the first of many company towns you'll see in the timber-rich Pacific Northwest. All the neat homes, with their nostalgic board sidewalks, are owned by the company, in this case Champion Paper. This particular mill produces plywood, so unlike the paper mills near Missoula, there is not a lot of talk about the effluents polluting the river. Just downstream from here is the famous Hellgate, one of the most strategic points in the geography of western Montana.

July 4th, 1806 [Lewis] the first 5 miles of our rout was through a part of the extensive plain in which we were encamped, we then entered the mountains with the East fork of Clark's river through a narrow confined pass on its N. side continuing up that river five Ms. further to the entrance of the Cokahlahishkit [Blackfeet] R.

Father De Smet, returning from a trip to his beloved Flatheads, camped here in 1846 and named it Porte de l'Enfer (Gate of Hell) because it was, as he put it, "the principal entrance by which parties of marauding Blackfeet reach the lands of our neophytes." The Mullan Road came through here; so did both the Northern Pacific and the Milwaukee railroads. Today, it's the canyon Interstate 90 uses to get from the wide valley near Deer Lodge to the great Missoula basin.

Missoula

Lewis spent his third Fourth of July in the wilderness here, and though the party had no swivel gun to fire in salute, nor a drop to fill a glass to toast the occasion, he liked what he saw.

July 3rd, 1806 [Lewis] as we had no other means of passing the river we buised ourselves collecting dry timber for the purpose of constructing rafts; timber being scarce we found considerable difficulty in procuring as much as made three small rafts. began to take over our baggage being obliged to return several times. the Indians swam over their horses and drew over their baggage in little basons of deer skins. by this time the raft by passing so frequently had fallen a considerable distance down the river to a rapid and was soon hurried down with the current a mile and a half before we made shore, on our approach to the shore the raft sunk. the vally of Clark's river is extensive beatifull level plains and praries. the tops of the hills and mountains on either hand are covered with long leafed pine larch and fir; near the river the bottoms are timbered with long leafed pine and cottonwood.

The river, which proved to be such an obstacle, you now can cross in about five seconds.

The city of Missoula has a population of only 33,000, but over 70,000 live in the county, making this the largest metropolitian center in western Montana. Though Missoula has its share of suburban shopping centers, the old downtown still has an active and somewhat cosmopolitan feel, thanks in part to the fact that this is a university town. Montana's flagship campus of its higher education system lies nestled at the base of the hills. Almost 9000 students are enrolled here. Downtown, you see fellows in pickups cruising up and down Orange Street on Saturday night, but you also can dine in exotic Greek, Mexican, Oriental, and vegetarian restaurants and browse in an old warehouse converted into boutique shops. The old **Greenough Mansion,** built in 1860, has been made into a restaurant. The new **Missoula Museum of the Arts** has changing exhibits and the **county courthouse** displays paintings by Montana artist Edgar Samuel Paxson. Paxson is most noted for his painting "Custer's Last Stand."

The Forest Service has its **Smoke Jumpers Center** at the airport, where you can see exhibits, motion pictures, and murals illustrating various fire-fighting techniques used in the forest. The old **Fort Missoula,** built during the Nez Perce War, is southeast of town. Fort Missoula was an internment camp for Italian POWs during World War II. There is a small museum on the base. Missoula and Salt Lake City are unique among American cities in that both have first-class ski resorts within a few miles of downtown.

The party was genuinely sorry to be leaving this country. Lewis's journal gives a hint about the affection he felt for the Nez Perce, and illustrates why he and Clark did so much to foster the new nation's relations with the American Indians.

July 2nd, 1806 [Lewis] I gave the Cheif a medal of the small size; he insisted on exhanging names with me according to their custom which was accordingly done and I was called Yo-me-kol-lick which interpreted is the white bearskin foalded.

July 3rd, 1806 [Lewis] I sent out the hunters who soon returned with three very fine deer of which I gave the indians half. These people now informed me that the road which they shewed me at no great distance from our Camp would lead us up the East branch of Clark's river and to a river they called Cokahlarishkit or the river of the road to buffalo and thence to medicine river and the falls of the Missouri where we wished to go. I directed the hunters to turn out early in the morning and indeavour to kill some more meat for these people whom I was unwilling to leave without giving them a good supply of provision after their having been so obliging as to conduct us through those tremendous mountains.

July 4th, 1806 [Lewis] at half after eleven the hunters returned from the chase unsuccessfull. I now ordered the horses saddled smoked a pipe with these friendly people and at noon bid them adieu. these affectionate people our guides betrayed every emmotion of unfeigned regret at separating from us

July 4th, 1806 [Gass] It is but justice to say that the whole nation to which they belong, are the most friendly, honest, and ingeneous people that we have seen in the course of our voyage and travels.

National Bison Range Thirty-seven miles north of Missoula is the 20,000-acre National Bison Range, operated by the U.S. Fish and Wildlife Service. This lovely park is at the southern end of the wide, beautiful Flathead Valley. Neither Lewis nor Clark ever saw this valley, so it's a bit of a detour off their trail, but a side trip well worth taking because it offers the best opportunity to see many of the animals that were common in 1805.

Buffalo live in India, bison live on the American prairie, you learn, but it seems as if we'll always call these massive, shaggy beasts buffalo. The range was established in 1908 with the help of President Theodore Roosevelt. The animals are mainly descendants of four young calves, brought here from the plains east of the divide in 1873 by a Pend d'Oreille (pronounced Pon-Der-Ray) warrior named Walking Coyote. The park rangers now maintain the herd at about 400 animals, the limit of what the land will support. A fine visitors center describes the life and habitat of the bison and tells the history of its exploitation. You can make a 15-minute drive to a nearby herd, but it's more rewarding to take the 2-hour self-guided tour through the entire park. The road is dirt and climbs steeply into several different wildlife habitats. At the higher elevations you have a good chance of seeing the pronghorn or antelope and the elk. A herd of bighorn sheep graze on the southern slopes. This is probably your best opportunity to see that magnificent animal. The Rocky Mountain goat also grazes in the park, but they're a bit reclusive and hard to spot. Other animals you have a reasonable chance of seeing include the black bear, long-tailed weasel, mink, skunk, coyote, bobcat, chipmunk, deer mouse, pack rat, and whitetail deer.

Travelers Rest

The spot Lewis and Clark called Travelers Rest is now virtually a suburb of Missoula. The Corps of Discovery returned here on June 30,

1806, after another harrowing journey over the rugged Bitterroots. The Flatheads were nowhere to be seen, and both captains worried that they had been massacred by the Gros Ventres, a tribe they called the "Minetarries of Fort de Prarie." The captains had previously decided to split the party here in order to have a better look around.

July 1st, 1806 [Clark] Capt. Lewis and Myself part at this place we make a division of our party and such baggage and provisions as is Souteable. the party who will accompany Capt. L. is G. Drewyer, Sergt. Gass, Jo. & R. Fields, Frazier & Werner, and Thompson Goodrich & McNear. I shall proceed on to the head of the Rockejhone with a party of 9 or 10 men.

At Travelers Rest we resume Lewis and Clark's westward trek. The Corps of Discovery is about to face its greatest challenge.

THE BITTERROOT MOUNTAINS

We are about to follow the Corps of Discovery during its worst time of crisis. It was getting late in the year, the explorers had already braved one snowstorm, water was freezing in their pots, and even in this low country they couldn't find enough to eat. Things were going to get a lot worse before they got better. To follow the route, drive west from Travelers Rest on U.S. Highway 12. Almost immediately you will come to an interesting spot.

On July 25, 1877, the Nez Perce came down out of the mountains with General Howard's army in slow pursuit. Here they encountered Captain Charles Rawn and 200 soldiers who had built a crude breastworks across the canyon. "We are going by you without fighting, if you will let us, but we are going by you anyhow," Chief Joseph is reported to have said. Two days of talks produced no results, so early on the morning of the 28th, the Nez Perce simply climbed up the mountainside and went around the canyon barricade. The place became known as **Fort Fizzle**.

Travelers Rest Creek

As you proceed westward, the canyon soon closes in all around. The journals give a hint of the trouble the captains were having.

September 11th, 1805 [Clark] A fair morning we set out at 3oClock and proceeded up on the Travelers rest Creek. about 7 miles Encamped at Some old Indian Lodges, nothing killed this evening hills on the right high & ruged, the mountains on the left high and Covered with Snow. The day Verry worm.

September 12th, 1805 [Clark] The road through this hilley Countery is verry bad passing over hills & thro' Steep hollows, over falling timber &c.&c. continued on & passed Some most intolerable road on the

Sides of the Steep Stoney mountains. Crossed a Mountain 8 miles with out water & encamped on a hill Side. our hunters Killed only one Pheasent this afternoon. Party and horses much fatigued.

A sign along the highway quotes Clark's entry of the 11th. The amazing thing is that, even in late summer, the mountain he was describing is still covered with snow. The canyon begins to widen out a bit; some hay is grown in the bottom lands, and the party came to one of the nicer spots they would encounter in the Bitterroots. It is clear, though, that they enjoyed the place more on their return the following spring.

Sept 13th, 1805 [Clark] I proceeded on with the partey up the Creek passed Several Springs which I observed the Deer Elk &c. had made roads to, and below one of the Indians had made a whole to bathe, I tasted this water and found it hot & not bad tasted.

June 29th, 1806 [Lewis] the prinsipal spring is about the temperature of the warmest baths used at the hot springs in Virginia. In this bath which had been prepared by the Indians by stoping the run with stone and gravel, I bathed and remained in 19 minutes, it was with dificulty I could reemain thus long and it caused a profuse sweat I observed that the indians after remaining in the hot bath as long as they could bear it, ran and plunged themselves into the creek the water of which is now as cold as ice can make it.

The **Lolo Hot Springs** are still in use, though they don't have the wilderness charm they did in 1806. Instead of a pool constructed from river stones and gravel, you now swim in a concrete plunge, surrounded by a chain-link fence. It's an old resort. The Thwaites edition of *The Lewis and Clark Journals* has a footnote reading: "There is now [1903] a good road up Lolo Creek to the springs and a daily stage-coach to and from Missoula."

Lolo Pass

That afternoon the party continued up Lolo Creek, crossed the pass, and dropped down into present-day Packer Meadows. U.S. Highway 12 more or less follows their route here, but keeps a bit to the north. You're in granite country now; the huge boulder on your right is what geologists call an erratic; it was pushed here by an advancing glacier. At the

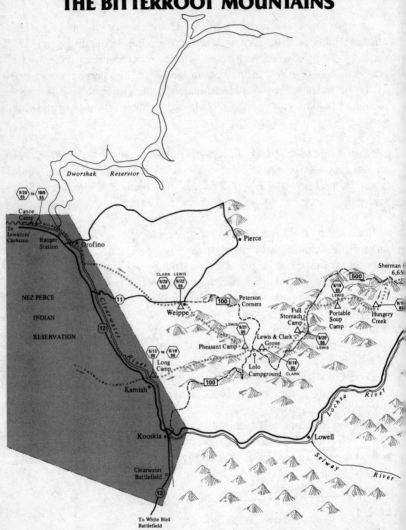

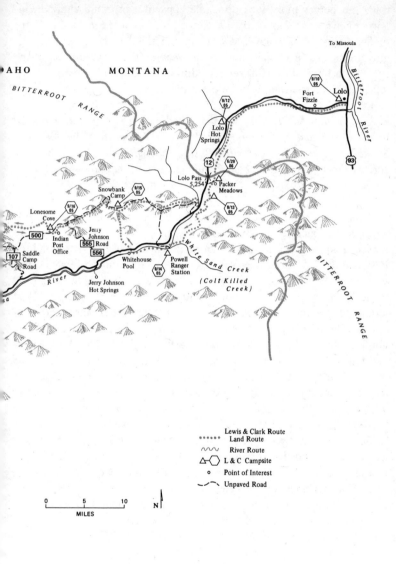

summit you cross into Idaho and enter the Pacific Time Zone. The Forest Service operates a **visitors center** here. A large-scale map shows the route our heroes took. Pick up a free copy of a brochure entitled *Following Lewis and Clark across the Clearwater National Forest*. The Corps of Discovery actually crossed Lolo Pass a mile to the east. Take the gravel road that leads past the visitor's center and park your car at the spectacularly beautiful **Packer Meadows**. A jeep road goes off to the left. It is a quarter-mile walk to the summit.

Sept 13th, 1805 [Clark] we proceeded over a mountain to the head of the Creek and at 6 miles from the place I nooned it, we fell on a Small Creek from the left which Passed through open glades. we proceeded down this Creek about 2 miles to where the mountains Closed on either Side crossing the Creek Several times & Encamped.

The Corps of Discovery had finally reached the summit of the Bitterroots, but once again fate was against them. They could hardly have picked a worse place to cross than here. Though the elevation of Lolo Pass is only 5254 feet, they would later have to go much higher. They were following an Indian road of sorts, but forest fires and "wind blows" made it hardly passable. Lt. John Mullan came through here in 1854 looking for a route to Walla Walla, and quickly decided to go elsewhere. The Nez Perce, as we just saw, crossed the pass in 1877. Then nothing happened. The railroad went north, along the Clark Fork River, and the highways followed. It wasn't until the 1930s that a motor road over Lolo Pass was considered, and then it took the Great Depression to get it built. The Civilian Conservation Corps constructed what it rather grandly called the Lolo Motorway, which more or less followed the route used by Lewis and Clark. The road never did live up to its imposing name. It never was paved, rarely opened before late July and closed two months later, and it was so rutty it took two days to traverse. Today it is known as Forest Service road 500. It still takes two days, and those with low-clearance automobiles are advised to avoid it altogether. Amazingly, it's only been in the last 20-odd years that the average traveler could drive to the pass that our heroes selected for their dash to the Pacific. U.S. Highway 12 wasn't completed until 1962.

You can still get a good idea of the type of country the Corps of Discovery had to traverse because at the summit it looks about the same now as it did then. Trees still lean this way and that and lie strewn about on the ground. Underbrush and fallen branches make passage difficult, even in this more or less level part of the trail. Packer Meadows, especially when the camas is in bloom, is still as pretty as it seemed to Meriwether Lewis in the spring of 1806.

June 29th, 1806 [Lewis] we ascended a very steep acclivity of a mountain about 2 miles and arrived at it's summit where we found the old road which we had pased as we went out. at noon we arrived at the quawmas flatts and halted to graize our horses and dine. we passed our encampment of the 13th September at 10 ms. where we halted there is a pretty little plain of about 50 acres.

Highway 12 drops quickly into the Lochsa River canyon. Within 4 miles you're below 4000 feet elevation. Lewis and Clark's trail joined the highway at the point where you first meet the river. Old Toby, their Shoshoni guide, should have led them immediately up the ridge on the north side of the road, but, inexplicably, he didn't. When the party returned in the spring that's the route they took, but now, when time was so precious, they went on a detour to the south. Apparently they wandered along on the southern flank of the canyon, at times going several miles up side streams before finally rejoining the river at present-day Powell Ranger Station. You can follow their 1806 return route by turning right on a dirt Forest Service road that climbs up the hill, eventually becoming the 500 road, the Lewis and Clark Motorway. The 500 road usually doesn't open until August, however, and since it is easy to get lost with so many logging roads criss-crossing the area, be sure to check in at the ranger station first.

Powell

U.S. Highway 12 passes by the **DeVoto Memorial Grove,** a lovely stand of cedars where the historian often camped when he was editing the Lewis and Clark journals. That was long before the highway came through here. A few miles farther on is Powell, the only settlement on the upper Lochsa River. The party camped here on the night of September 14. It was only their fourth night in the wilderness, but already things were getting scary.

Sept 14th, 1805 [Clark] Encamped opposit a Small Island at the mouth of a branch on the right side of the river which is at this place 80 yards wide, Swift and Stoney, here we were compelled to kill a Colt for our men & Selves to eat for the want of meat & we named the South fork Colt killed Creek, the Mountains we passed to day much worst than yes-

terday the last excessively bad & thickly Strowed with falling timber. our men and horses much fatigued.

Captain Lewis must have been having some second thoughts about his boast on August 20, when he was still with the Shoshoni: "If the Indians can do it, we can do it."

The Powell Ranger Station is in a lovely grassy meadow a mile or so off the highway. Nearby is the Powell Resort, a delightful spot in the middle of the wilderness. Guests stay in log cabins that have no running water, but dine in country splendor in the timbered main building. Fall is the big season for this resort; hunters book rooms a year in advance. There is a gas pump out front, something you wouldn't normally care about, but it's the only one for another 70 miles.

Four miles below Powell you come to **Whitehouse Pond,** named after Private Whitehouse, who was the only one to mention the pond in his journal. Many of the place names along the Lolo Trail were promulgated by Ralph Space, retired superintendent of Clearwater National Forest and a Lewis and Clark scholar. It was here that the captains learned old Toby had led them astray. These western canyons are much too V-shaped to be usable for primitive roads. That's why it took so long for this highway to be built. You either take to the river, something not possible in a dugout cannoe, or head for the ridges. The explorers had dropped down to 3000 feet. The only way out of this abyss was to climb to 7000 feet—2000 feet higher, it turned out, than they were at Lolo Pass. This day would test their mettle.

September 15th, 1805 [Clark] We set out early and proceeded on over Steep points rockey & buschey as usial for 4 miles. here the road leaves the river and assends a mountain winding in every direction to get up the Steep assents & to pass the emence quantity of falling timber. Several horses Sliped and roled down Steep hills which hurt them verry much. after two hours delay we proceeded on up the mountain Steep & ruged as usial. when we arrived at the top we could find no water and Concluded to Camp and make use of the Snow to cook the remns. of our Colt & make our Super. Two of our horses gave out, pore and too much hurt to proceed

Snowbank Camp, as it is now called, is located along the 500 road. Things were going from bad to worse.

Sept 16th, 1805 [Clark] began to Snow about 3 hours before Day and continued all day. at 12 oClock we halted on the top of the mountain to worm & dry our Selves a little. I have been wet and as cold in every part as I ever was in my life, indeed I was at one time fearful my feet would freeze in the thin Mockirsons which I wore, we Encamped in a thickly timbered bottom which was scurcely large enough for us to lie leavil, men all wet cold and hungary. Killed a Second Colt

On this day the Corps of Discovery attained the greatest altitude in the Bitterroots—7033 feet—near a spot known as **Indian Post Office.** Their campsite on this, their sixth night in the wilderness, is at **Lonesome Cove,** so named because Private Whitehouse wrote: "We descended the mountain down in a lonesome cove on a creek where we camped."

All these events, keep in mind, happened up on the ridge north of Highway 12. The paved road keeps to the river all the way through the Bitterroots. If you're willing to brave the backcountry, the steep, 10-mile Jerry Johnson road, Forest Service number 565/566, goes up the hill to Lonesome Cove, which is alongside the 500 road. This and another road 15 miles farther down the highway, #107 or the Saddle Camp Road, provide the two principal access roads to Lewis and Clark's actual route. A loop trip connecting the two roads takes about 4 hours, but again, check with the rangers at Powell or, if you are coming the other way, at Orofino, before going off into the backwoods. Also, be especially wary of logging trucks. Eighteen-wheelers come off this mountain, and the drivers aren't used to having to cope with tourists.

A couple of miles below the Jerry Johnson road a trail goes south to a hot springs with the same name. Park at the footbridge, cross the river, and take the trail to the right. It's a beautiful mile-long hike through Douglas fir and cedar forests to the springs, which look much like what Lewis and Clark saw at Lolo. Hikers, like the Indians before them, have arranged the rocks to form little bathtub-sized pools.

Meanwhile, the explorers were struggling along, up on the ridge. Their next night's camp was at a place called **Sinque Hole,** located a couple of miles west of where Saddle Camp Road, number 107, meets the 500 road. The Forest Service rangers say this is the single best place to get a look at the high country along the actual Lewis and Clark Trail.

September 17th, 1805 [Clark] Cloudy morning our horses much Scattered which detained us untill one oClock. Snow falling from the trees which kept us wet all the after noon. Killed a fiew Pheasents which was not sufficient for our Supper which compelled us to kill Something, a Coalt being the most useless part of our Stock he fell a Prey to our appetites. we made only 10 miles to day.

The next day things were getting so desperate Lewis decided on a bold plan. Someone would have to go ahead and summon help.

Sept 18th, 1805 [Lewis] Cap Clark set out this morning to go a head with six hunters. ther being no game in these mountains we concluded it would be better for one of us to take the hunters and hurry on to the leavel country a head and there hunt and provide some provisions while the other remained with and brought on the party. this morning we finished the remainder of our last coult. we dined & suped on a skant proportion of portable soupe, a few canesters of which, a little bears oil and about 20 lbs. of candles form our stock of provision.

So, on the eighth night in this terrible country, they dined on candles. The thick forest prevented them from seeing ahead, so they had no idea how much farther these awful mountains would extend. Clark, though, now ahead of the main party, got a glimmer of hope.

September 18th, 1805 [Clark] we passed over a countrey Similar to the one of yesterday more fallen timber. from the top of a high part of the mountain at 20 miles I had a view of an emence Plain and leavel Countrey to the S W. & West. Encamped on a bold running Creek which I call Hungery *Creek as at that place we had nothing to eate.*

The Corps of Discovery had been on this ridge for four days, and even with horses they had come only about 40 miles and still were nearly 7000 feet above sea level. The next morning the main party saw the prairie, too.

September 19th, 1805 [Lewis] Set out this morning a litle after sun rise and continued our route for 6 miles when the ridge terminated and we to our inexpressable joy discovered a large tract of Prairie country lying to the S.W. through that plain the Indian informed us that the Columbia river, (in which we were in surch) run. this plain appeared to be about 60 Miles distant, but our guide assured us that we should reach it's borders tomorrow. the appearance of this country, our only hope for subsistance greately revived the sperits of the party already reduced and much weakened for the want of food.

Unfortunately, once again old Toby was wrong. Their ordeal wouldn't end "tomorrow," and they still had "today" to get through.

September 19th, 1805 [Lewis] the road was excessively dangerous being a narrow rockey path generally on the side of a steep precipice, Fraziers horse fell from this road and roled with his load near a hundred yards into the Creek. when the load was taken off him he arose to his feet & appeared to be but little injured. we encamped in a little raviene, haveing traveled 18 miles. we took a small quantity of portable soup, and retire to rest much fatiegued. several of the men are unwell of the disentary.

Lewis's character shows through most in this time of crisis. The next day, on the 20th, as cold, wet, hungry, and miserable as he must have been, his journal opens with a description of a new bird ("it's note is ch a-ah-cha-ah"), comments on the quality of the soil (dark gray), and closes with notes on the observance of alder, huckleberry, and a new kind of honeysuckle.

The day before Captain Clark had come upon a stray Indian horse, which he shot, butchered, and left the bulk of the carcass here for the main party. The following day, Lewis's eleventh in the wilderness, proved to be his nadir.

September 21st, 1805 [Lewis] we killed a few Pheasants, and I killed a prarie woolf [coyote] which together with the ballance of our horse beef and some crawfish which we obtained in the creek enabled us to make one more hearty meal, not knowing where the next was to be found. I find myself growing weak for the want of food and most of the men complain of a similiar deficiency, and have fallen off very much.

It's difficult to follow the route of these last few days. The 500 road goes one way, the party went another. It's best, therefore, to leave the intrepid explorers for a bit, continue on down to civilization, and then pick up their trail on the Weippe Prairie as they came straggling out of the mountains.

Kooskia As you proceed on down the Lochsa, you notice a gradual change in the flora. Later blooming wild flowers begin to appear and you spot an occasional deciduous tree along the river. This is a "catch and release" stream—anglers are prohibited from using bait, and all fish must be returned. The practice, while not helpful to the larder, greatly increases the sport of trout, steelhead, and salmon fishing. There are numerous campgrounds all along this road. The first bit of civilization you come to is at the tiny town of **Lowell,** where there is a cafe, gas pump, and several small motels. This is the confluence with the Selway, which means "smooth water" in the Nez Perce language. Lochsa means "rough water," so the rough and the smooth combine to form the middle fork of the Clearwater, which Lewis and Clark called the Kooskooskee. In spite of its name, the Selway has the reputation for being one of the most challenging whitewater rivers in the country. The elevation here is under 2000 feet, the lowest you've been since leaving North Dakota.

Highway 12 enters the **Nez Perce Indian Reservation** at the town of Kooskia, situated at the confluence of the South and Middle forks of the Clearwater River. With a population of 800, this is the first real settlement you've seen since leaving Montana. You turn right here to rejoin the Lewis and Clark trail, but consider making a detour 40 miles south to visit the places where the tragic Nez Perce War began.

White Bird Battlefield State Highway 13 goes up the south fork of the Clearwater. About 5 miles south of Kooskia is a sign that offers comments on what became known as the Clearwater Battle. On July 11, 1877, General O. O. Howard's lumbering army crossed the river, hoping to take the Nez Perce by surprise. His hopes came to naught and the battle came to an end with the Nez Perce withdrawing to the east.

The highway soon leaves the river and climbs up on the **Camas Prairie,** a lovely wheat-growing area where the tribes used to come to camp, dance, sing songs, and collect the camas root, a staple of the Nez Perce, Walla Walla, and Cayuse diets. The Corps of Discovery traveled over this prairie on its return from the ocean.

> *May 7th, 1806 [Lewis] the face of the country when you once have ascended the river hills is perfectly level and partially covered with the long-leafed pine. the soil is dark rich loam thickly covered with grass and herbatious plants which afford a delightful pasture for horses. In short it is a beautifull fertile and picturesque country.*

The Nez Perce war began in this "beautifull fertile and picturesque country." These people, whom the captains described as "among the most amiable men we have seen," felt they had been pushed to the limit. On the morning of June 13 three young bucks, angered at being forced onto the reservation and seeking revenge for a white man's murder of one of their fathers, rode out on a bloody raid. By the end of the second day about 15 white settlers lay dead. Knowing the army would retaliate, the Nez Perce retreated south, where they laid a trap.

The town of **Grangeville,** population 4000, is the commercial center of the Camas Plateau. Here you pick up U.S. 95, Idaho's principal north-south highway. At the **White Bird Summit,** elevation 4245 feet, you have a beautiful view of the grassy canyon 2500 feet below. A new road, built to interstate standards, drops into this Snake River tributary, but to tour the battlefield take the old highway, which twists its way down in a series of lovely switchbacks. The National Park Service sells a self-guided auto tour brochure, which describes the details of what happened on June 17: Displaying all the skill and courage that would take them almost to Canada, the Nez Perce, outnumbered two to one, put to rout a batallion of troops led by Captain David Perry. The score: US—34 dead, Nez Perce—0.

Kamiah

To continue following Lewis and Clark's trail, return to Kooskia, turn north on U.S. Highway 12, and proceed 6 miles to the lumber town of Kamiah. To the Nez Perce, this is one of their most important places. Nearby, a rock outcropping is the Heart of the Monster, where, their mythology says, the nation was created. The Monster who lived here was killed and dismembered by Coyote, an important figure in Nez Perce legend. As Coyote flung the pieces to the other parts of the land, different peoples sprang up: Coeur d'Alene, Cayuse, Pend Oreilles, Flathead, Blackfeet, Crow, Sioux. When he finished he realized that he had made no people for the land where he stood, so he took the monster's heart and squeezed the blood out of it. These drops mingled with the earth and became the Ne-Mee-Poo, the Nez Perces. This and other lovely legends are retold in *Coyote Was Going There: Indian Literature of the Oregon Country.*

On the return trip Lewis and Clark arrived here in early May—too early, as it turned out.

May 7th, 1806 [Lewis] The Spurs of the Rocky Mountains were perfectly covered with snow. the Indians inform us that the snow is yet so deep that we shall not be able to pass them untill the next full moon; others set the time at still a more distant period. this is unwelcom intelligence to men confined to a diet of horsebeef and roots, and who are as anxious as we are to return to the fat plains of the Missouri

The "unwelcom" news was, unfortunately, correct. The party remained here until June 10, and even then they had to pause later on because they were still too early to brave the Bitterroots. Though the going this time, much of it over hard-packed snow, was much easier than on their westward trek, more than two months would pass before they reached "the fat plains of the Missouri." Kamiah became known as "The Long Camp." Although they were no doubt disappointed, they evidently had a pretty easy time of it, for the journals are sprinkled with comments like these:

May 13th–June 10th, 1806 [Biddle] It is an extensive level bottom, thinly covered with long-leaved pine, with a rich soil and broken hills on the east and northeast, with the best game in the neighborhood . . . they [Nez Perce] soon gave us two fat young horses, without asking anything in return, an act of liberal hospitality much greater than any we have witnessed since crossing the Rocky mountains . . . Their chief subsistence is roots, and the noise made by the women in pounding them gives the hearer the idea of a nail factory . . . afterward the hunters killed a female bear with two cubs . . . The air is pure and dry; the climate is quite as mild as, if not milder than that of the same parallels of latitude in the Atlantic States . . . this district affords many advantages to settlers, and if properly cultivated, would yield every object necessary for the subsistence and comfort of civilized man . . .

Nez Perce hospitality was returned as well:

May 11th, 1806 [Lewis] many of the natives apply to us for medical aid which we gave them cheerfully so far as our skill and store of medicine would enable us. schrofela, ulsers, rheumatism, soar eyes, and the loss of the uce of their limbs are the most common cases among them.

Weippe

To rejoin Lewis and Clark's 1805 trail proceed northward on U.S. Highway 12 until you come to a road junction at Greer. State Highway 11 branches off to the right, crosses the river, and climbs steeply to the Weippe (pronounced WE-ipe) Prairie. The canyon down below is fairly heavily timbered with white pine. Since you're now heading back toward the Bitterroots, you expect to see more forest, so it's a great shock to find that you suddenly arrive in the midst of wheat fields. The elevation here is about 3000 feet. The Nez Perce had lived on this prairie for 8000 years. Until September 20, 1805, not one of them had ever seen a white man.

September 20th, 1805 [Clark] I set out early and proceeded on through a Countrey as ruged as usial passed over a low mountain and at 12 miles decended to a level pine Countrey. proceeded on through a butifull Countrey for three miles to a Small Plain in which I found maney Indian lodges. at the distance of 1 mile from the lodges, I met 3 Indian boys, when they saw me they ran and hid themselves in the grass, I desmounted gave me gun and horse to one of the men [and] gave them Small pieces of ribin & Sent them forward to the village. Soon after a man Came out to meet me, with great caution & Conducted me to a large Spacious Lodge They call themselves Cho pun-nish or Pierced noses. the fiew men that were left in the Village and great numbers of women geathered around me with much apparent signs of fear. those people gave us a Small piece of Buffalow meat, Some dried Salmon beries & roots all of which we eate hartily. I find myself verry unwell all the evening from eateing the fish & roots too freely

This historic event occurred at the hamlet of Weippe, located on the eastern end of the prairie, bordering on the forest. It's a lovely little out-of-the-way village, with a grade school, grocery store, restaurant, and, of course, a place where the locals can gather, the Elk Horn Bar. Several small, family-owned, red cedar and Douglas fir sawmills provide employment for those fortunate enough to live here.

The day after Clark arrived in Weippe, he sent the hunters out for game while he worked at making friends with the Nez Perce. The hunters spent most of the day at their task, but were unsuccessful:

September 21st, 1805 [Clark] I purchased as much Provisions as I could with what fiew things I chanced to have in my Pockets, Such as Salmon Bread roots & berries, & Sent R. Fields with an Indian to meet Capt. Lewis

By the next day, the ordeal of the Corp of Discovery was over. Reuben Fields found the main party about 15 miles east of Weippe.

Sept 22nd, 1805 [Lewis] we met Reubin Fields whom Capt Clark had dispatched to meet us with some dryed fish and roots that he had procured from a band of Indians, I ordered the party to halt for the purpose of taking some refreshment. I divided the fish roots and buries, and was happy to find a sufficiency to satisfy compleatly all our appetites. the pleasure I now felt in having tryumphed over the rockey Mountains and decending once more to a level and fertile country where there was every rational hope of finding a comfortable subsistence for myself and party can be more readily conceived than expressed, nor was the flattering prospect of the final success of the expedition less pleasing.

A Forest Service campground is located at the spot where the Corps of Discovery spent its last night in the wilderness. It's a 16-mile drive over mostly unpaved but nevertheless good roads, and you'll travel through some spectacularly beautiful country. Drive due east out of Weippe, jog north for a bit, and then turn south at Peterson Corners. From there, follow the 100 Road to its junction with Road 500 at the Lolo Campground. The Lewis and Clark grove is nearby. Beyond the campground, Road 100 descends into the Clearwater Canyon, where it meets U.S. Highway 12 near Kooskia. This road makes a nice loop trip because you don't have to retrace your steps back to Weippe.

The captains were now reunited on the Weippe Plateau.

September 22nd, 1805 [Biddle] As we approached the village most of the women, though apprised of our being expected, fled with their children into the neighboring woods. The men, however, received us without any apprehension, and gave us a plentiful supply of provisions. The plains were now crowded with Indians, who come to see the persons of the whites, and the strange things they brought with them

Meanwhile, Captain Clark, who had befriended the Nez Perce Chief Twisted Hair, was anxious to learn about the country that lay ahead.

September 22nd, 1805 [Clark] I got the Twisted hare to draw the river from his Camp down which he did with great Cherfullness on a white Elk skin, from the 1st fork which is few miles below, to the large fork on which the So So ne [Snake] Indians fish, is South 2 Sleeps; to a large river which falls in on the N W. Side [Columbia] and into which the Clarks river *empties itself; is 5 Sleeps from the mouth of that river to the* falls; *is 5 Sleeps at the falls he places Establishments of white people &c.*

So, the first map ever made of the upper Columbia River was drawn here at Weippe, on an elkskin pelt. Chief Twisted Hair's story proved reasonably correct, except there were no white people at the (Celilo) falls.

Pierce

Gold was discovered at the nearby town of Pierce in 1860. The discovery set the stage for the Nez Perce War. Governor Isaac Stevens of

Washington Territory had previously signed a treaty with the tribe, essentially giving them their homeland intact. The Nez Perce had no use for the yellow metal—blue beads were prettier—but the white man did, and the rush was on. By 1863 the old treaty was abrogated and a new one signed, one which essentially gave the Nez Perce a tenth of the land they had previously held. Land-grabs like this festered in the Nez Perce's minds for nearly 15 years. It's not hard to imagine why those warriors on the Camas Prairie went on their rampage in 1877.

Pierce, named for the man who found the gold, is a bit larger than Weippe; it even has a motel, which, during the fall, is filled with hunters. There are a few mementos of the gold rush around, but mostly this is timber country. A large sawmill is nearby. According to a sign along the road, Pierce had some wild days in its youth, including a time when it, too, had to resort to law according to the vigilante committee. A paved but poorly graded road leads north and then west, where it drops down to Orofino, the largest town in the Bitterroots.

Orofino

The expedition emerged from the high country a mile or so upstream from Orofino, a little bit worse for the wear. The change in diet, from buffalo to nothing and then to dried fish and camas roots, caused the entire party to come down with stomach disorders and diarrhea.

September 24th, 1805 [Clark] at Sunset we arrived at the Island on which I found the Twisted hare, *and formed a Camp. Capt Lewis scercely able to ride on a jentle horse which was furnished by the Chief, Several men So unwell that they were Compelled to lie on the Side of the road for Some time others obliged to be put on horses. I gave rushes Pills to the Sick ths evening.*

Dr. Rush's pills were a diuretic. Scholars think there was a bacteria in the dried salmon to which the Nez Perce had developed an immunity, but which affected the newcomers greatly.

Orofino is a quiet town with lovely brick buildings that are being refurbished. The word of E. D. Pierce's discovery quickly spread, and by 1861 this became a hell-roaring mining camp. The prospectors who flooded into town erected wooden frames that they covered with muslin. These were known as "muslin houses" and the streets on which they stood were called "Muslin Row." The poet and sometime judge Joaquin Miller is said to have worked as a Pony Express rider between here and the low country. Nowadays, the red-brick county courthouse and other restored buildings give the place a real mountain-west feel. Several cafes, bars, and motels cater mostly to the sportsmen who come to this lovely country to hunt and fish. The headquarters of the **Clearwater National Forest** is just west of town, where you can get detailed maps of the area and advice about the condition of local roads. Nearby is the massive concrete **Dworshak dam,** one of the highest in the United States. The reservoir backs up water on the North fork of the Clearwater for 50 miles, and has become a popular recreation area.

Canoe Camp

Lewis and Clark spent some time at the confluence of the North and Main forks of the Clearwater, querying the Nez Perce before concluding that this was where they should build canoes and take to the river. Today, the correctness of that decision seems obvious, but keep in mind that they were exploring unknown territory. Six years later Wilson Price Hunt's party of Astorians faced the same quandary, but failed to check out what lay ahead. That trip, so colorfully detailed by Washington Irving in *Astoria,* proved to be a disaster. Once again we see just how wilderness-wise our two leaders proved to be.

Those who were not so sick they couldn't move set to work building canoes. Lewis was impressed with the size of the trees. While still back in the mountains he wrote: "I saw several sticks today large enough to form eligant perogues of at least 45 feet in length."

October 3rd, 1805 [Gass] All the men are now able to work; but the greater number are very weak. To save them from hard labour, we have adopted the Indian method of burning out the canoes. The morning of the 7th was pleasant, and we put the last of our canoes into the water; loaded them, and found that they carried all our baggage with convenience. We had four large canoes; and one small one, to look ahead.

With this comment, we modern-day Lewis and Clark trekkers can leave the primitive beauty of the Bitterroots behind and also look ahead, toward what is now called the Inland Empire.

THE INLAND EMPIRE

After branding their horses and leaving their saddles in a cache near the river, the party set off downstream. The character of the country changes rapidly here. Within a few miles the pines begin to disappear, first from the lowlands near the river and then on the higher slopes as well. We have just moved into a completely different climatic zone, much drier and almost treeless.

October 7th, 1805 [Biddle] For some miles the hills are steep and the low grounds narrow; then succeeds an open country with a few trees scattered along the river We passed in the course of the day ten rapids.

Clearwater River

The Corps of Discovery soon found itself on waters quite unlike anything they had seen before. Although they had learned the skills of running rivers like the Allegheny and the Ohio, they weren't prepared for the rigors of these western torrents. They had become seasoned mountain men, but they had not yet perfected their downriver skills in dugout canoes. It's not surprising; they had been traveling for a year and a half and this is the first time they found a river going their way.

October 8th, 1805 [Clark] passed 15 rapids just below, one canoe in which Sergt. Gass was Stearing was nearle turning over, she Sprung a leak or Split open and Bottom filled with water & Sunk. the men, Several of which Could not Swim hung on to the Canoe, I had one of the other Canoes unloaded & with the assistance of our Small Canoe and one Indian Canoe took out every thing & toed the empty Canoe on Shore.

The experience must have been particularly unsettling for Sergeant Gass. Water never seemed to agree with Patrick. He lived chiefly on whiskey after he got home, yet he survived to see his 99th year. The

THE INLAND EMPIRE

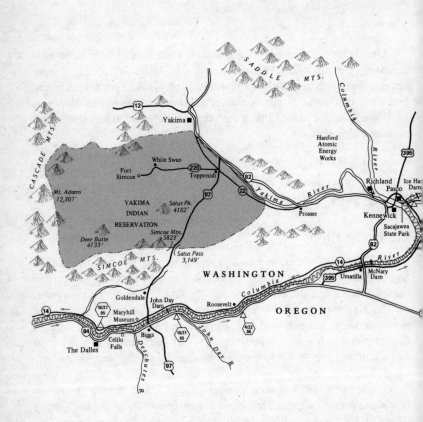

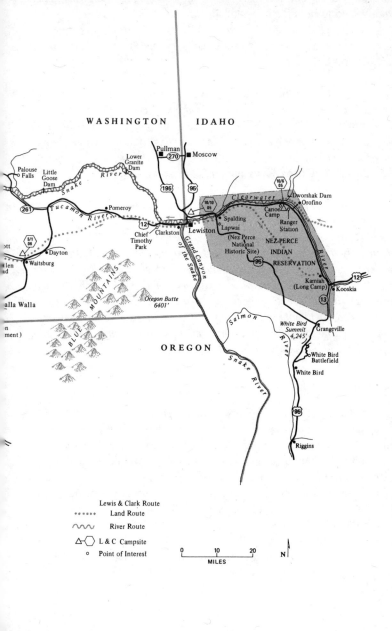

party learned quickly. By the time they reached the Columbia gorge they were shooting rapids even the natives wouldn't dare, though it's unlikely they ever achieved the skills of the French-Canadian fur trappers, who proved to be the masters of this art.

The changed eating habits were still giving them trouble, but that's nothing new; today's jet-setters suffer the same malady, though the modern cure is a bit different.

October 10th, 1805 [Biddle] Being again reduced to fish and roots, we made an experiment to vary our food by purchasing a few dogs, and after having been accustomed to horse-flesh, felt no disrelish for this new dish. The [Perce Nez] have great numbers of dogs, which they employ for domestic purposes, but never eat; and our using the flesh of that animal soon brought us into ridicule as dog-eaters.

They received the same scorn on their return trip:

May 5th, 1806 [Lewis] while at dinner an indian fellow verry impertinently threw a poor half starved puppy nearly into my plait by way of derision for our eating dogs and laughed very heartily at his own impertinence; I was so provoked at his insolence that I caught the puppy and threw it with great violence at him and struck him in the breast and face, siezed my tomahawk and shewed him by signs I would tommahawk him, the fellow withdrew apparently much mortifyed and I continued my repast on dog without further molestation.

Spalding

The Corps of Discovery now neared Spalding on the Nez Perce Indian Reservation. The village, with its old general store, is the site of an 1836 Presbyterian mission, Idaho's first. The National Park Service has built a fine interpretive center on a grassy bluff above the river. The exhibits don't dwell much on Lewis and Clark, though there is a replica of the canoe the explorers used. Most of the displays are designed to show the culture, art, and living style of the Nez Perce. The Nez Perce were of Salishan linguistic stock, a family that also includes the Cayuse, Umatilla, Wallawalla, and Klickitat tribes. The interpretive center shares with the one at the Yakima Nation the honor of being the most comprehensive and sensitive Native American exhibit west of the Continental Divide.

Lewis and Clark's journals have nothing but good things to say about the Nez Perce.

May 12th, 1806 [Clark] The Indians brought up a fat horse and requested us to kill and eate it as they had nothing else to offer us to eate.

May 17th, 1806 [Biddle] The [Nez Perce] are in general stout, well formed, and active. [They] are among the most amiable men we have seen. Their character is placid and gentle, rarely moved to passion, yet not often enlivened by gayety.

Seventy-one years later the Nez Perce put the lie to that last statement. **Fort Lapwai** had been built near here, and in May 1877 commander General Oliver Otis Howard summoned chiefs Joseph, White Bird, and Looking Glass to a powwow. They were accompanied by Joseph's brother Ollokot, and the medicine man Toohoolhoolzote. The message was simple: Quit your villages in the Wallowa Valley and move onto the reservation. These so called "nontreaty" Indians had never been party to the 1863 agreement, and stoutly maintained that the pro-American, Christian leader who signed it, a man named Lawyer, had no right to speak for the rest of the tribe. Howard would have none of that argument; he had Toohoolhoolzote jailed for a while so he wouldn't have to listen to the great debater's arguments. By the time Joseph and his brethren had returned to the Wallowa soldiers were already occupying the valley. Joseph sadly moved his tipis, women, children, and livestock to the Camas Prairie where, as we have seen, the war soon began.

Fort Lapwai was one of the first missions in the Pacific Northwest. Reverend Henry Spalding and his wife Eliza came west with a party of missionaries in 1836 and originally settled near Walla Walla with the Whitman family. Encouraged by the Indians' apparent embracement of Christianity, he and his wife moved here in 1838. Along with religion they taught reading, writing, agriculture, and home economics. Fort Lapwai became the site of Idaho's first school and church, and early on, it housed a gristmill and printing press. Old Joseph, Chief Joseph's father, was baptized here in 1840. The same thing happened here that we have seen all along the Lewis and Clark Trail. The Nez Perce were not really interested in the cross. They wanted "medicine," meaning magic, magic which would make their crops grow, their women fertile, their children healthy, and, above all, make them more powerful than their enemies. They became discouraged when Christianity didn't live up to their ex-

pectations. Reverend Spalding and his wife fled for a time after the
Whitmans were massacred, but they returned, and he continued to work
among the Nez Perce until his death in 1874.

About 1500 Nez Perce live at Fort Lapwai now. The agency headquarters is a few miles south of the visitors center, where the **Pi-Nee-Waus Community Center** is located. A tour of the town leaves you feeling as if we're no closer to solving the Indians' problems than we ever have been. A 12-foot-high chain-link fence surrounds the high school, making it look like a prison. The government-built houses look as if they belong in an army camp rather than out in this beautiful countryside.

Lewiston/Clarkston

Ten miles farther west is the confluence of the Clearwater and Snake rivers, site of the twin cities that honor our heroes. One of the ironies of the expedition is that 40 miles back they could hardly move forward because there were too many downed trees. Now they found themselves in country with nary a stick to build a fire.

October 10th, 1805 [Biddle] The country at the junction of the two rivers [Clearwater and Snake] is an open plain on all sides, broken toward the left by a distant ridge of high land, thinly covered with timber. This is the only body of timber which the country possesses; for at the forks there is not a tree to be seen. This southern branch [Snake] is in fact the main stream of Lewis' [Salmon] river, on which we camped when among the Shoshonees. The Indians inform us that it is navigable for 60 miles further up.

The river is navigable, in a sense. No steamboat ever got very far upstream, though two did make it down from Ft. Boise at the height of a spring flood. Other potentially adventuresome captains looked at these battered hulls and wisely decided to wait for a better technology. That technology appeared with the advent of the jet boat. A sizable number of outfitters run boats up what is now called "Hells Canyon." At one point, near He Devil Peak, Idaho, the treeless mountains soar 7900 feet above the river to form America's deepest gorge, a good 2350 feet deeper than Arizona's Grand Canyon. The rapids aren't what they used to be because of upstream dams, but there's still a bit of whitewater to get your adrenaline going. Half-day and one-day tours are offered, as well as longer overnight excursions, where you can stay in a lodge accessible only by boat or camp out under a tipi. The outfitters claim excellent bass, trout, and, in season, steelhead fishing.

The confluence of the Snake and Clearwater is the present head of navigation for the thousands of barges that ply the Columbia River watershed. Lewiston is a port city, strange as that may seem. Look across the river at the dock areas. You'll see the expected grain elevators, but also stacks of containers bearing the names of some of the biggest shipping companies in the world, Hapag-Lloyd and K-Lines to name but two. In the old days stern-wheeled steamboats used to tow the barges upriver. The last survivor of a fleet of nearly 600 is now moored a few miles upriver. The *Jean,* built in Portland, was as modern as they came when she was launched in 1938. With a steel hull and superstructure, and weighing in at more that 500 tons, she sported a compound engine, supplied by steam from an oil-fired boiler. Though she looks like a conventional stern-wheeler, there are actually two wheels in back, each of which can be separately driven, forward or back. Her builders bragged:

> Set one wheel turning forward and the other astern, twitch the rudder, and the towboat turns nicely in her length without going forward at all. Or turn the rudder hard over, and the Jean *begins to crab sideways directly abeam . . . the* Jean, *with her accomplishments, is a mighty handy steamer when the channel is swift and the moorage tight.*

The old beauty stayed in service until the late-1940s. She is now the property of the State of Idaho. The hope is that enough money can be raised through donations to keep her in operating condition.

The combined population of Lewiston, Idaho, and Clarkston, Washington, is about 35,000. Lewiston, by far the larger, is a company town; the Potlach Corporation owns the large paper mill along the Clearwater. The elevation here is only 700 feet, making this the lowest town in Idaho and, interestingly, lower than St. Louis, Missouri. Lewiston has had a bit of urban renewal, although it's hard to think of a town this small as urban. The grand old **Lewis and Clark Hotel** has been converted into an office complex, and nearby, some old brick warehouse buildings have been turned into a shopping mall. The main street is dotted with interesting shops and restaurants, and the numerous bait, tackle, and gun stores remind you that this is vacation country. The bank buildings provide the most interesting modern architecture. The Corps of Engineers has built a park called **Tsceminicum,** meaning "Meeting-of-the-Waters," at the confluence of the Clearwater and Snake rivers. A large sculpture interprets Nez Perce mythology with a symbolic Earth Mother figure "whose

body sustains all forms of life, and from whose hands the rivers run."
On one side, her flowing hair forms images of life and legend on the
Snake, on the other the Clearwater.

Clarkston, by contrast, has a bit more western-cowboy feel. The most
popular spot in town is a combination newsstand, gift store, coffee shop,
and card room. Walk in any morning and you'll likely see 30 or 40 men
(and some ladies) fingering their cards and rattling their chips. Nightlife
is about the same in both towns, the liveliest spot being the one where
the country-music band is playing on a given night of the week. Cultures seem to mix in these parts; tonight's action might be in a Chinese
restaurant.

Snake River

Lewis and Clark wouldn't recognize today's river below Lewiston.
Between here and Bonneville, the Corps of Engineers has constructed
so many dams that there is no free-running river left. Our route, U.S.
Highway 12, follows their westward journey for a few miles along the
shore of a lake created by the Lower Granite Dam. The canyon begins
to close in near **Chief Timothy Park,** where the State of Washington
has built the **Alpowai Interpretive Center.** Exhibits and audio-visual
programs depict the geologic and human history of the area.

On their return from the Pacific, Lewis and Clark were able to procure enough horses to allow them to abandon their canoes and proceed
overland. Highway 12 more or less follows that route between here and
the village of *Waitsburg.* The country continues to be dry and rather
bleak until you cross the Alpowa summit, where you descend into the
beautiful Tucamon Valley. Here you encounter the first substantial
farmland since leaving Montana. *Pomeroy,* population 1700, looks much
like many of the towns we visited in Kansas and Nebraska, with farm
implement dealers lining the roads coming into town.

Palouse Falls

Twenty miles west of Pomeroy the highway turns south and begins
to climb out of the Tucamon Valley. To pick up a portion of Lewis and
Clark's downstream voyage, turn right on State Route 261, which goes
to one of the more remote portions of the Snake River. Vermillion cliffs,
covered in the spring by patches of green stubgrass, make an enchanting
sight. Half a dozen campers are parked by the river, their owners no
doubt out in skiffs, trolling for bass. Highway 261 follows the river for

a bit, crosses on a new bridge, and then climbs steeply to a plateau on the north. This is the famous Palouse country.

> *April 25th, 1806 [Lewis] the river hills are about 250 feet high and generally abrupt and craggey in many places faced with a perpendicular and solid rock. level plains extend themselves from the tops of the river hills to a great distance on either side of the river. the soil is not as fertile as about the falls, tho' it produces a low grass on which the horses feed very conveniently. I did not see a single horse which could be deemed poor and many of them were as fat as seals.*

The horses Lewis referred to are quite likely the Appaloosa, which are native to this area. Horses were introduced to North America by the Spaniards in the 18th century. Unlike other plains tribes, the Nez Perce selectively bred their stock from animals that had drifted up from Mexico. The Nez Perce horses were called Palouse, which the fur traders corrupted to Palousey, which in due time was corrupted again to Appaloosa. The name Palouse now refers to the wheat-growing country between here and Spokane. The soil is rich, consisting mainly of volcanic ash carried here by the winds from the volcanoes to the west. There has never been a drought in the Palouse; dry farming techniques produce very high yields. Most people think of Washington as a timber state, but it also produces 10% of all the wheat grown in the United States.

The Corps of Discovery never saw it, but there is an impressive waterfall up one of the side canyons. Highway 261 crosses the river and climbs into cattle country, as wild and scenic as anything Montana or Wyoming has to offer. The range riders here aren't the pretentious type, and they're too far from civilization to put on airs. The fellow you see on horseback is as likely to be wearing a woollen stocking cap as he is a ten-gallon hat. Five miles beyond the river, turn right onto a dirt road that heads out into the sagebrush. There's a water trough here, and you'll probably have to stop while a white-face or two struggles to its feet to get out of the way. The mile-long dirt road descends to a mesa overlooking the Palouse gorge, where the river drops several hundred feet in one mighty cascade. Washington State has built a nice campground here.

Dayton

Return now to U.S. Highway 12, which turns south, climbs over a low divide, and drops into the Touchet River Valley. Lewis and Clark called this Drouillard's River, honoring the great hunter, but the name didn't stick. This is the land of the Wallawalla, a nation that was particularly kind to the explorers on their return trip.

> *April 29, 1806 [Biddle] In the course of the day we gave small medals to two inferior chiefs, each of whom made us a present of a fine horse. We have, indeed, been treated by these people with an unusual degree of kindness and civility.*

As we will shortly see, this respect between the two vastly different cultures lasted a scant 40 years.

The principal town in the Touchet Valley is Dayton, population 2500, where not only wheat but apples, sugar beets, and vegetables are grown. Green Giant has a packing plant here. The railroad arrived in 1887. A station built in the gingerbread style of the day has been lovingly restored by the townspeople. The building contains period furniture as well as office equipment, telegraph apparatus, and the signaling devices of that nostalgic era.

The State of Washington built a lovely little roadside park just west of Dayton that they named after our heroes. Another couple of miles brings you to the charming town of **Waitsburg,** where U.S. 12 turns south. To follow Lewis and Clark's trail more closely, though, drive through town and go west on route 124, the highway that goes directly to the Tri-Cities area, a place we will visit a little later on. The farmland along here is as pretty as anything in Iowa, and just as rich. Ten miles west of Waitsburg, turn south on Highway 125. You've now rejoined our old friend the Mullan Road, which we followed over the Continental Divide west of Great Falls, Montana. By the time Lt. Mullan's construction crews got here, they were only 18 miles from their objective, Walla Walla.

Walla Walla

"A town so lovely they named it twice," the natives say of their gemlike city in the midst of this fertile valley. Walla Walla has a farmtown feeling, with a plethora of farm equipment dealers, feed and fertilizer stores, and the ubiquitous grain elevator. The town is also the home of Washington State Prison, but nobody talks much about that. What they do talk about, especially at events like Kiwanis Club meetings, is how to keep their town nice. You can't cut a tree down without someone's approval. The result is that this city, unlike many these days, has managed to keep its downtown vital. Instead of building a suburban shopping center, they restored the lovely turn-of-the-century brick buildings on Main Street. Unfortunately, however, the nine-story red-brick **Marcus Whitman Hotel** has been turned into condos, so you can't stay in what was once the finest hotel in the area. Walla Walla even has an extensive local bus system, something unusual for a town of only 26,000. There's no big-town feel here, but rather a sense of community. Daybreak finds the parking places in front of the **Red Apple Cafe** crowded with pickup trucks. It's a social time for the farmers, who tablehop agiley, exchanging gossip as adroitly as urbanites at a Manhattan cocktail party.

Walla Walla's showpiece is **Whitman College,** the oldest school in Washington. Founded in 1859, it now boasts a student body that comes from all over the country. Ivy-covered brick buildings blend with newer ones of concrete, all framed by century-old elm trees. The nearby city park with its Victorian bandstand reflects a similar graciousness.

The army built a fort here in 1859 and named it after the first **Fort Walla Walla,** which was located 25 miles west on the Columbia River. A Veterans Administration hospital now occupies the site. A museum on the south end of the hospital grounds depicts much of the history of the area. About 20 restored buildings give a sense of life in the late 1800s.

Proceeding westward on U.S. Highway 12, you follow the route of the first railroad in the Inland Empire, which ran between Walla Walla and the Columbia River. Built by the ingenious Oregon pioneer, Dr. Dorsey S. Baker, it was very much a homemade road. The rails were wood, covered with iron straps. The rolling stock was of the crudest type, including a locomotive imported from Pittsburgh in 1872. The line was variously called "Dr. Baker's road," "the strap iron road," and the "rawhide road." During the gold rush the good doctor charged $6 per ton for the 36-mile run, which in a short time made him wealthy. You're now passing through the rich farm country where they grow the "Walla Walla sweet," quite possibly the best onion in the world. Ten miles west is Waiilatpu.

The Whitman Mission at Waiilatpu By the mid 1830s tales of the glories of the west were beginning to filter back east. The Lewis and Clark journals had been published, and the Astorians had returned to St. Louis with tales of a Shangri-la in Oregon. Some Nez Perce had visited St. Louis, seeking the secret to the white man's medicine. After the trappers, the first Americans to answer the call of the west were the missionaries. In 1936 Dr. Chester Maxey, then president of Whitman College, speculated on why they came.

Inspired by motives as varied and mixed as the manifestations of religion can be—utter self-immolation, emotional impulses of varied complexity, the urge for social service, the call of the benighted, the fascination of the unfamiliar, the exigencies of sectarian competition—a tidal movement of missionary endeavor swept in upon the Oregon country in the early 30's.

One of the men to answer the call was Dr. Marcus Whitman who, with his wife Narcissa and several other missionaries, arrived in the Walla Walla Valley in 1836. They named their mission Waiilatpu, "the place of the rye grass." For 11 years Whitman and his followers farmed the area, provided a haven for Oregon-bound travelers, and tried to convert "the aborigines who stubbornly refused to be reclaimed from heathendom and savagery." The Wallawalla and Cayuse tribes were becoming more and more upset about the incursions of the whites, when, to again quote Dr. Maxey:

. . . an epidemic of measles, a paleface pestilence unknown to the older Indians, swept the valley and claimed many victims . . . to the primitive's mind this was clearly a result of the white doctor's evil magic. The more Dr. Whitman did to treat the sick, the more the conviction spread that he was killing them with some subtle poison.

The culmination of this episode was the massacre on November 29, 1847, of Dr. Whitman, the lovely and articulate Narcissa, and 12 other men. Of the 50 who were captured, 3 children subsequently died, and the rest were ransomed for 62 blankets, 63 cotton shirts, 37 pounds of tobacco, and a certain quantity of guns and ammunition. Some of the women had been compelled to "marry" their Cayuse captors. Another 20 years would elapse before settlers again felt safe enough to venture into this land that Lewis and Clark had found so hospitable.

The mission site is now administered by the National Park Service. When the exhibits were first set up by the local inhabitants of the valley, there was an unabashed attempt to glorify the great doctor. "St. Marcus of the Rye Grass," he came to be called by those who thought the Cayuse should get better shrift. Today American attitudes are changing, brought about by a better understanding of the fact that atrocities were perpetrated by both the natives and the whites. The events at Waiilatpu were clearly a clash of two cultures, and the exhibits planned for the renovated center will no doubt cast a slightly different perspective on the reasons for the disaster.

Wallula

Highway 12 meets the Columbia, where we again join the westward route of the Corps of Discovery.

October 18th, 1805 [Clark] passed an Island on which was 2 Lodges of Indians drying fish on Scaffolds. the river passes into the range of high Countrey, at which place the rocks project into the river. at the Commencement of this high country a Small riverlet falls in which appears to have passed under the high country in its whole course. Saw a mountain bearing S.W. conocal form Covered with Snow.

The mountain is Mt. Hood, which Lt. William R. Broughton of the Vancouver expedition of 1792 named for a prominent English admiral of the day. The Corps of Discovery first passed this spot in the fall when the river was low. On their return trip, they called the Walla Walla "a handsome river." Though there's nothing here now save a truckstop cafe, this was an important place in the winning of the west. The next white people to pass by were the half-starved Wilson Price Hunt party, who arrived here in January 1812. But it was a Canadian, Donald McKenzie, who established the first settlement. This intrepid fur trader worked variously for the Northwest Company, John Jacob Astor's Pacific Fur Company, and finally for the Hudson's Bay Company. In 1818 he built a trading post at this spot. Originally called Fort Nez Perce, it became the first Fort Walla Walla, the only center of civilization between Fort Vancouver and the great American plains. Gardens were started, cattle brought in to feed on the meadowlands, and industry was developed to supply the needs of agriculture. Jedediah Smith, perhaps the greatest pathfinder and trailmaker of all time, paused here. So did Nathaniel Wyeth and B. L. E. Bonneville, both of whom tried to break the Hudson's Bay Company's monopoly on the fur trade, but were thwarted when the factors at Fort Walla Walla refused to provide them with anything, save enough food to get out of the territory. The existence of the fort played a major role in Marcus Whitman's decision to establish his mission nearby.

The fort was abandoned in 1856 when the Hudson's Bay Company retreated to Canada, and any vestige of the building was lost in the flood of 1894. Finally, the closing of the gates of the new **McNary Dam** obliterated even the lonely site of this once famous outpost. A few miles to the north is a huge Boise Cascade paper mill where cardboard containers are manufactured.

The Tri-Cities

U.S. Highway 12 turns north and follows the Columbia upstream to its confluence with the Snake. When the explorers finally reached the long sought Columbia, they seemed quite unmoved. Their writings contain no hurrah, no whoopee, no "Oh the joy," a phrase they used when they first saw the Pacific. By their calculations they had come 3714 miles.

October 16th, 1805 [Clark] after getting Safely over the rapid and haveing taken Diner Set out and proceeded on Seven miles to the junction of this river and the Columbia which joins from the N.W. We halted above the point on the river to smoke with the Indians who had collected there in great numbers to view us, great quantities of a kind of prickley pares, much worst than any I have before seen.

The State of Washington has built an interpretive center at **Sacajawea Park,** on the exact spot where the party camped. The building has many excellent exhibits, including a private collector's roomful of Native American artifacts and displays that emphasize the kinds of tools these tribes used.

After crossing the Snake River you enter **Pasco,** the first of three towns that make up a metropolitan area of 90,000 people. For the first time since leaving Omaha you feel a bit as if you're in a city; the area seems laced with freeways. Pasco, founded in 1884, is a railroad and industrial center. The two main streets, Lewis and Clark, have become largely deserted; most of the merchants have moved out to the suburban shopping center near Richland. Across the Columbia is another city with no downtown: **Kennewick,** which means "Winter Haven." A beautifully designed suspension bridge connects it with Pasco. The largest of the Tri-City towns, **Richland,** is also the newest. Built by the government as part of the Manhattan Project in the 1940s, it is the home of the Hanford Atomic Energy Works. The plutonium used for the bomb dropped on Nagasaki was processed here. The Department of Energy operates a **science museum** near the center of town. Most exhibits are aimed at the high-school level of understanding. Washington Public Power Administration's ill-fated nuclear power plants are located a few miles north of town.

Much of the Inland Empire has an arid feel. The annual rainfall is only eight inches, but thanks to water diverted from the Columbia at Grand Coulee Dam, a great deal of land is now planted with the kind of crops you find in Iowa and Missouri. You see the same giant sprinkler systems irrigating fields of corn, alfalfa, and soybeans. Farmers are planting vineyards and apple orchards, and farther west, in the Yakima Valley, you see huge fields of hops. The city of Yakima is known as the apple center of the nation.

Side Trip to Toppenish It's a bit off the Lewis and Clark trail, but the **Yakima Nation Heritage Center** is an interesting place to visit. Take Interstate 82 west from Richland, turn off at the town of Prosser, cross the Yakima River, and continue westward on State Highway 22 to Toppenish, a name that means "people from the foot of the hills." This road is a bit more rural than the freeway, which it parallels for about 40 miles. Ahead looms the great snow-clad dome of Mount Adams, which Captain Clark mistook for another volcano farther west.

October 19th, 1805 [Clark] I descovered a high mountain of emence hight covered with Snow, this must be one of the mountains laid down by Vancouver, as seen from the mouth of the Columbia River. I take it to be Mt. St. Helens,

The Cultural Center, located near the town of Toppenish, is unique in that it is run by the Yakimas themselves, the only museum of its kind. Dioramas and exhibits tell the story of the Yakima people and their culture. There is a gift shop, library, and a huge restaurant that features buffalo steaks and Columbia River salmon as well as traditional American dishes. Other nearby sites of interest are **Fort Simcoe,** a military post established in 1856, and the reservation village of **White Swan,** where powwows and rodeos are held several times during the summer. The city of *Yakima,* population 17,000 is 30 miles farther west.

In 1805 the natives, most likely Yakimas, whom Lewis and Clark called the Sokulks and Chimnapums, gave the Corps of Discovery a cordial welcome.

October 16th, 1805 [Clark] The Chiefs then returned with the men to their camp; Soon after we purchased for our Provisions Seven dogs, Some fiew of these people made us presents of fish. The 2 old chiefs precured us Some fuil Such as the Stalks of weeds or plants and willow bushes. one man made me a present of about 20 lb. of verry fat Dried horse meat.

October 17th, 1805 [Clark] Those people appeare of mild disposition and friendly disposed. Their ammusements are similar to those of the missouri. they are not beggerlly, and receive what is given them with much joy.

The Yakima Nation Heritage Center has a showcase displaying a drawing of this meeting and a sign that reads:

One day downriver from the East came strangers, white men and a black man in canoes. We received them as honored guests, as tradition would have us do. In a few days they departed downstream to the West. Life for our people would never be the same.

The next exhibit compares the lands the Yakima had in 1805 with the size of the reservation today.

U.S. Highway 97 south from Toppenish climbs quickly into some very pretty pine-forested mountains. After crossing 3100-foot Satus Pass you drop into the nice little town of **Goldendale,** where an old mansion has been converted into an historical museum. The highway south of town climbs over a little hillock where you can see all four of the Columbia River volcanoes: Mt. Hood is 50 miles south; Adams, 45 miles to the west; St. Helens, 25 miles beyond that; and Mt. Ranier lies 85 miles to the northwest. You rejoin the Lewis and Clark trail at Maryhill.

Columbia Plateau

To rejoin the Lewis and Clark trail from the Tri-City area go south on Interstate 82, which crosses the Columbia River near Umatilla. Then take the lonely but well-constructed State Route 14. The river here forms a deep cleft in the otherwise flat Columbia plateau. Once you drop down to water level things seem rather bleak. The lack of any buildings and the abundance of sand and sagebrush make you feel as if you're really out in the desert. Actually, you're passing through incredibly rich wheat land. There's a whole hidden empire up on the plateau. Your first hint of its value is when you see the giant grain elevator in the town of **Roosevelt,** population 93. That grain is coming from somewhere. A little later you spot a dirt road heading up the hill. A sign reads CIRCLE R RANCH. SALESMEN BY APPOINTMENT ONLY. We all know that salesmen don't waste their time in a desert. This is dry-farming country. Tractors, huge air-conditioned behemoths with eight 6-foot-high tires can till a swath 40 feet wide.

As Lewis and Clark proceeded down the river, they suddenly found the natives more difficult to deal with. The historian Bernard DeVoto speculates that the lower Columbia tribes had been debauched by the fur traders. The explorers encountered one man wearing a sailor's coat, so there is no doubt that they had previously met the white man.

October 21st, 1805 [Clark] those people did not receive us at first with the same cordiality as those above.

Near Roosevelt the homeward-bound explorers, now quite out of goods to trade with, quit the river and went overland. They had been buying horses for some days and felt they needed more.

April 22nd, 1806 [Lewis] we obtained 4 dogs and as much wood as answered our purposes on moderate terms. we can only afford ourselves one fire, and are obliged to lie without shelter, the nights are cold and days warm.

April 24th, 1806 [Lewis] the natives had tantalized us with an exchange of horses for our canoes, but when they found that we had made our arrangements to travel by land they would give us nothing for them. I determined to cut them in pieces sooner than leave them on those terms, Drewyer struck one of the canoes and split off a small piece with his tommahawk, they discovered us determined on this subject and offered us several strands of beads for each which were accepted.

The beads, no doubt, came in handy for trading later on.

Near the junction with U.S. Highway 97 you pass **John Day Dam,** named for one of the Astorians who came west with the Wilson Price Hunt party of 1812. Poor John Day. He was a fine man from Kentucky who barely kept his sanity on their ill-fated westward trip. When, in the return trek, it looked like he was in for more of the same, he went completely insane and had to be sent back to the coast, where he died shortly thereafter. Most of the power from John Day Dam is used by the nearby aluminum reduction plant.

Maryhill

On September 8, 1883, near Gold Creek, Montana, the last spike was driven on the Northern Pacific Railroad, thereby linking the Columbia River with the east by a band of steel. Joining the Northern Pacific at Spokane was the Spokane, Portland & Seattle Railroad, which boasted that it followed the "water level route" to the Pacific. James J. Hill, the "Empire Builder," acquired the railroad and with it much of the land along the right bank of the Columbia. He probably would have turned over in his grave if he'd known what his son-in-law, Sam Hill, did with some of that property. Sam built his dream house high on a bluff overlooking the river and the prairie. He soon realized he could not get his wife to live there, so he set out to convert it into a museum. Help came from some unlikely places. The San Francisco sugar heiress, Alma de Bretville Spreckles, got interested in the project. So did Queen Marie of Romania. Between the three, they amassed an incredible collection of art, which is here for us to appreciate today. A hundred miles from the nearest city, you're treated to an impressive Rodin collection, jewels and mementos from the Romanian court, old chess sets, some fine paintings, a collection of dolls depicting the Paris fashions of the 1940s, and an extensive display of Native American artifacts. The site of the museum is spectacular, framed on the southwest by the beautiful Mt. Hood and below by the mighty Columbia. Golden wheat fields surround the green gardens of the castlelike estate.

Nearby is another one of Sam Hill's creations, a concrete replica of the Druid ruins at Stonehenge. Sam improved on things here: He added the stones that have been missing from the site at Devonshire since the dawn of history.

Biggs

Across the river from Maryhill, near Biggs, Oregon, the Oregon Trail finally rejoins the route of Lewis and Clark. A plaque beside the road reads, "Between 1843 and 1863, at a spot not far from here, westbound emigrants caught their first sight of the Columbia River." By land, we're 1500 miles from St. Joseph, where we last encountered the Oregon Trail. The Corps of Discovery traveled a bit farther: 2100 miles, not counting numerous wriggles in the river. Both Interstate 84 and the Union Pacific Railroad parallel the Oregon Trail. By some accounts, upward of half a million immigrants started the trek and as many as 30,000 were lost along the way. A small state park, located just off the interstate, is at a place where the immigrants had a tough time crossing the Deschutes River. At this, our first stop in Oregon, it's appropriate to comment that it is pronounced, Or-re-gun, with an equal accent on all the syllables, never Or-re-GONE.

Celilo Falls

Celilo, now nothing more than a wide spot in the road, was in 1805 what one might call the Bosporus of the Pacific Northwest, the place where east meets west. Celilo Falls, the most impressive cascade on the Columbia, was by far the best place to fish for salmon. Whichever tribe controlled the falls also controlled the commerce. Scholars estimate that at the time of Lewis and Clark, 125 different tribes lived in the Pacific Northwest, and they spoke 56 different languages. The seminomadic inland tribes, Klickitat, Wishram, and Wasco, would come downstream to trade with the stay-at-home Clackamas, Wahkiakums, and Chinook. George Vancouver, when he put in at Gray's Harbor in 1792, noted that in spite of the differing linguistic stock, the tribes had developed a common language used only for trade. He called it "Chinook Jargon."

Before the white man arrived on the scene the natives had developed a sort of modus operandi for getting past the falls.

April 11th, 1806 [Biddle] These people do not, as we are compelled to do, drag their canoes up the rapids, but leave them at the head as they descend, and carrying their goods across the portage, hire or borrow others from the people below.

The tribes, possibly realizing that with the Corps of Discovery there would be no quid pro quo, and not yet having learned the art of toll-collecting, resorted to thievery:

April 11th, 1806 [Biddle] The labor of crossing would have been very inconvenient if the Indians had not assisted us in carrying some of the heavy articles on their horses; but for this service they repaid themselves so adroitly that, on reaching the foot of the rapids, we formed a camp in a position which might secure us from the pilfering of the natives, which we apprehend much more than we do their hostilities. Three men were left behind to guard the baggage. This precaution was absolutely necessary to protect it from the Wahchellahs, whom we discovered to be great thieves. They crowded about us while we were taking up the boats, and one of them had the insolence to throw stones down the bank at two of our men. We now found it necessary to depart from our mild and pacific course of conduct.

By 1843, when the settlers began to arrive, the tribes had learned the white man's ways. The tolls charged became so high that Sam Barlow found it profitable to build an alternate route to the Willamette Valley. During the height of the steamboat era several daring captains ran the rapids downstream on the spring freshet, emerging with battered but still seaworthy hulls. A portage railroad was built in the 1860s, but it was not until 1905 that the Corps of Engineers finally constructed a lock system around the cataract. Boats could then steam all the way from the tidewater to Lewiston, Idaho.

The falls were majestic. A member of the Astor party, Russ Cox, wrote:

> *Above this channel for four or five miles the river is one deep rapid, at the upper end of which a large mass of high black rock stretches across from the north side and nearly joins a similar mass on the south. They are divided by a strait not exceeding fifty yards wide. Through this narrow channel, for upwards of half a mile, the immense waters of the Columbia are one mass of foam, and force their headlong course with a frightful impetuosity, which cannot at any time be contemplated without producing a painful giddiness.*

The natives built platforms out over the base of the falls where, during the salmon run, they could stand and scoop out the fish with a long-handled net. This practice continued into the late 1950s. One old-timer boasted that the fish were so numerous you could walk across the river on their backs without getting your moccasins wet. Now there's nothing but a lake; the falls were innundated by the closing of the gates at the Dalles Dam.

For six miles below Celilo Falls the explorers came on one rapid after another. William Clark's journal sounds like an ad for a modern-day white-water rafting outfitter.

> *October 23rd, 1805 [Clark] The whole of the Current of this great river must at all Stages pass thro' this narrow chanel of 45 yards wide. as the portage of our canoes over this high rock would be impossible with our Strength, and the only damage in passing thro thos narrows was the whorls and swills arriseing from the Compression of the water, and which I thought by good Stearing we could pass down Safe, accordingly I deturmined to pass through this place notwithstanding the horrid appearance of this agitated gut swelling, boiling & whorling in every direction, which from the top of the rock did not appear as bad as when I was in it; however we passed Safe to the astonishment of all the Inds. who viewed us from the top of the rock.*

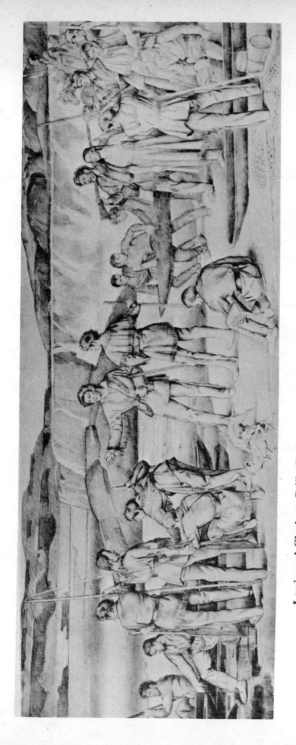

Lewis and Clark at Celilo Falls: This mural, which adorns the Oregon Capitol Rotunda, shows the explorers portaging their canoes around a major Columbia River cataract. Sacagawea is at the center right, York is sitting at the far right. Courtesy of the Oregon Historical Society.

Horsethief State Park, on the Washington side of the river, is near the spot where the Corps of Discovery demonstrated such splendid whitewater skills. The natives carved a number of pictographs into the soft sandstone, including one charming drawing of a person's head that is called "She Who Watches."

The Dalles

The Dalles rhymes with gals. The word comes from the French, meaning flagstone or stepping stone, which no doubt refers to the basaltic outcroppings that used to puncture the water along this part of the river. The Dalles was more or less the end of the Oregon Trail, for the cliffs just below town at Rowena essentially blocked further overland travel. A thriving business was established to float wagons down the river on rafts, which soon made The Dalles the center of commerce for the entire area. It was to the white man what Celilo Falls had been to the natives. Daniel Lee established a Methodist mission here in 1836. The army built a post in 1847 that was enlarged to a full-fledged fort during the Indian Wars of 1850–1853. Parts of the fort are still extant. Follow the signs up the hill to a building called the Surgeon's Quarters, which has been turned into a nice museum. Also located on the grounds is a collection of Oregon Trail–type wagons. Another museum is in the old courthouse located at the west end of downtown.

The Dalles is on the upper end of what became the middle river, the relatively placid 35-mile stretch of water between here and Cascade Falls. As such, it became an important shipping port. The Union Pacific Railroad has a large marshaling yard here. If you look carefully along the tracks east of town, you might see a depression-style hobo jungle. The Dalles has a giant plywood mill and the last of the grain elevators you will see on this odyssey. You've come to another of those transition zones where, in the next few miles, the countryside changes dramatically. The average annual rainfall in The Dalles is 14 inches. Thirty-five miles away it's 65 inches. You're about to leave the cowboy west and the Inland Empire behind. Ahead lie the forests of the Cascade Mountains.

THE CASCADES

The Cascades section of the Lewis and Clark trail is only 60 miles long, but in that short distance there is so much to see and do, you could well spend a week poking around in this spectacularly beautiful country. We pick up the Corps of Discovery as it is about to leave terra incognita. It's perhaps wise to pause for a moment and consider what Lewis and Clark knew about the country ahead.

Exactly three centuries after Columbus, Captain Robert Gray, an American, sailed up the coast of North America. He was lucky. The Englishman George Vancouver was in the area, too, and both were looking for the fabled Northwest Passage. Vancouver, by studying the water and the currents along the coast, knew that a great river discharged into the ocean at this latitude. Unfavorable tides and winds, however, prevented him from entering the river. Captain Gray came along a few weeks later and found things going his way. Having successfully crossed the bar, he anchored at present-day Gray's Bay, Washington, where he promptly claimed the country for America and named the river Columbia, after his ship. That same summer Vancouver sent his second in command, Lieutenant William Broughton, back to the area. Broughton sailed nearly 150 miles upstream and, in the process, named Mount Hood for a prominent admiral and St. Helens after the British Ambassador to Spain, Baron St. Helens. Both these voyages were well chronicled and therefore known to the learned men along the Delaware and Potomac, including Jefferson and his private secretary, Meriwether Lewis. As we saw earlier, William Clark's journal noted seeing "Mt. St. Helenas, laid down by Vancouver." The explorers were expecting to find a range of mountains ahead, and of course they did.

October 29th, 1805 [Clark] after brackfast we proceeded on, the mountains are high on each side, containing scattering pine white Oake & under groth, hill Sides Steep and rockey: Here the mountains are high

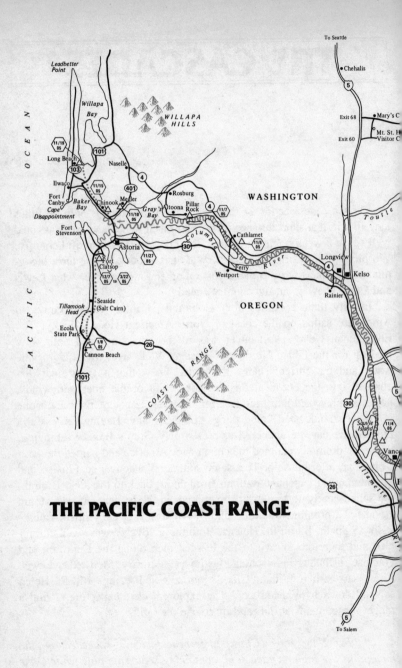

THE PACIFIC COAST RANGE

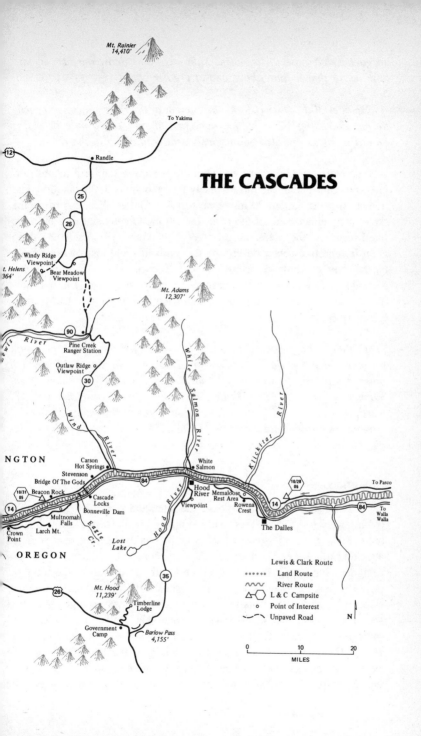

on each side, those to the Lard. Side has Some Snow on them at this time, more timber than above and of greater variety.

October 30th, 1805 [Clark] the current of the river is also verry jentle not exceeding 1½ miles pr. hour and about ¾ of a mile in width. Stones or rocks are also permiscuisly scattered about in the river,

The Corps of Discovery was about to leave the semiarid Inland Empire and enter one of the most densely forested areas in the country. The transition occurs about 20 miles west of The Dalles. You can make the drive along either bank of the river, though the Oregon side is a bit more interesting. At the Memaloose Safety Rest Area on Interstate 84 there is an intrepretive center describing the hardships and loss of life caused by the river's turbulence prior to the construction of the dams and locks. Memaloose (a Klickitat name meaning "Land of the Dead") Island, now mostly submerged, was the site of an ancient burial ground that the captains explored.

October 31st, 1805 [Clark] I saw 8 vaults for the Dead. the bones in some wer 4 feet thick, in others the Dead was yet layed side of each other nearly East & west wraped up & bound securely in robes, great numbers of trinkets Brass Kittles, sea shells, Iron Pan Hare &c. &c. were hung about and great maney wooden gods, or Imeges of men cut in wood, set up around the vaults. I can not learn certainly if those people worship those wooden emiges.

Rowena Crest

A more scenic route is to leave the interstate at exit 84 and take the old highway to Mosier. This road, completed in 1916, was immediately dubbed "a poem in stone," such was its beauty. Rather than brutally attacking this rugged country, as the engineers of the interstate did, the designer, Samuel C. Lancaster, chose to adapt his road to the topography. By ducking in and out of side canyons and climbing hills with great sweeping curves, he created a masterpiece in highway design. Only two sections are still extant, 16 miles between The Dalles and Mosier and the even more beautiful 12-mile section between Bonneville and Sandy River. Here, the two-lane road rises gently to the Rowena Crest, the barrier that blocked further land passage on the Oregon Trail. Park your car at the paved overlook and stroll out among the wild flowers. The view from this plateau, especially in the spring, is lovely. The brown hills on the Washington side become forested above a thousand feet or so, and then are capped by green pastures, dotted with barns and farmhouses. The scene looks much like a Swiss alp. Snowy Mt. Adams looms above the river, and below you can see towboats trailing chevron ripples in the placid water. A meadowlark sings from a distant bush, while a Canada goose wings his way toward an unseen marsh. The explorers observed the birds, too:

> *October 30th, 1805 [Clark] Saw Some fiew of the large Buzzard. Capt. Lewis Shot at one, those Buzzards are much larger than any other of ther Spece or the largest Eagle white under part of their wings &c.*

> *November 2nd, 1805 [Clark] saw great numbers of waterfowl of Different kinds, such as Swan, Geese, white & grey brants, ducks of various kinds, Guls, & Pleaver.*

The buzzards were, in fact, the California Condor, which has the largest wingspan of any bird in North America. They also first identified the spoonbill or shoveler duck.

Hood River

After dropping down from the plateau, the backroad passes through the first of many apple orchards located on the eastern side of the Oregon Cascades. Hood River, population 16,000, is a farming town, as evidenced by the large packing house along the railroad. The Corps of Engineers built a park and small boat launching facility along the lake formed by Bonneville Dam. A new shopping center and resort motel are nearby.

The crown jewel of Hood River is the **Columbia Gorge Hotel,** which sits on a steep bluff a mile west of town. This lovely country inn, built in 1921 by the lumber baron Simon Benson, resembles an Italian villa. The 15-foot ceiling in the entry hall sets the mood. So does the safe behind the front desk, which seems as big as a house. Sofas nestle around a cheery fire, and the antique-adorned dining room looks out over country as pretty as anything you have seen on this entire trip. The highlight of your stay is breakfast, where you are served a dozen kinds of fresh fruit and melons, oatmeal, three eggs, bacon, ham, sausage, honey with rolls, and pancakes. That's not a choice, mind you; they bring all of the above. The place should be called the Columbia Gorge-Yourself Hotel.

Those with more modest tastes can stay in any of a half dozen motels on both sides of the river. Of special interest is Washington State's Carson Hot Springs, whose buildings seem little changed since the resort was built in 1897. The 126° mineral spring is said to help in the cure of arthritis, skin afflictions, poison oak, kidney and internal disorders, neuritis, rheumatism, and stomach disorders.

Mount Hood and the Barlow Road

For a spectacular view of the Cascades, go south on State Highway 35 for a mile or so and then turn left on Panorama Road. You climb steeply for a couple of miles and then enter apple country. The road swings around a hill and then climbs to its summit where, all of a sudden, you see Mt. Hood, looming majestically above the orchards. If you're lucky enough to be here on a sunny spring morning when the trees are in bloom, you'll get the photo of a lifetime. Highway 35 continues south, winding through the orchards for about 15 miles, and then begins to climb into the forested country. This highway, like most in Oregon, is well built and well maintained. After another 15 miles you come to 4600-foot Bennett Pass, drop a little, and then climb to the famous Barlow Pass. In 1846 Captain Samuel J. Barlow, incensed with the outrageous tolls being charged for portage on the Columbia, organized a company to build a road over the south flank of Mt. Hood. It was completed early enough to enable nearly 150 immigrant wagons and 1500 head of livestock to pass over it during the fall of that year. It remained a toll road until 1912. Much of the route parallels U.S. Highway 26, the road from here to Portland, which you join just south of the pass.

A few miles to the west, turn right on a side road and make the steep climb to **Timberline Lodge,** one of the few good legacies of the Great Depression. This masterpiece of mountain architecture was constructed by hundreds of artisans and craftsmen employed by the Civilian Conservation Corps in the late 1930s. Franklin Roosevelt came here personally to make the dedication speech. Though she's showing her age, Timberline Lodge is a marvelous place to stay. Rooms are tucked under turrets and beneath soaring roofs. The sitting room, five stories high, is dominated by a central fireplace as big as a small house. A bar on the balcony looks out over a snow field with a chair lift that skiers use year-round. The large dining room has an old-resort feel, with brightly dressed young skiers doing the serving chores.

White Salmon

White Salmon, population 2000, is directly across the Columbia from Hood River. A toll bridge replaced the ferry that used to cross here. The short but spectacular White Salmon River is a playground for fishermen and white-water enthusiasts. During the summer, daily 3-hour raft trips are made down the boiling cascade. The put-in point is 12 miles north of the town of White Salmon. The Corps of Discovery, of course, didn't have the inflatable rubber rafts that make white-water running not only possible, but for the most part quite safe. They did, however, have the good fortune to improve the quality of their watercraft.

October 23rd, 1805 [Clark] I obvserved on the beach near the Indian Lodges two butifull canoes of different Shape & Size to what we had Seen above wide in the middle and tapering to each end, on the bow curious figures were cut in the wood &c. Capt. Lewis exchanged our Smallest canoe for one of them by giveing a Hatchet & few trinkets to the owner who informed that he purchased it of a white man below for a horse, these canoes are neeter made than any I have ever Seen and calculated to ride the waves, and carry emence burthens

The trade probably saved the day for, as we shall see, the explorers desperately needed seaworthy boats when they got down to the wind-swept waters near the ocean. Just west of White Salmon is a relic of a bygone era. Look up the hillside and you'll see a flume where rough-sawn logs, called "cants," are floated 9 miles down from a mill up on the Little White Salmon River. The Broughton flume was built in the 1920s and is still considered to be the most practical way to get timber from the forest to this mill located along the railroad tracks.

The Columbia Gorge

Below Hood River, the countryside changes even more rapidly. Geologists believe the gorge was formed in recent times, perhaps no more than 700 years ago, by a giant slide. The dry-soil-loving yellow pines soon dissappear, replaced by Douglas fir, the most important forest product in the Pacific Northwest. Underbrush becomes thicker and ferns and moss begin to appear. A sign at a trailhead challenges hikers to see if they can identify the 50 species of plants along the path. The explorers were happy to once again be in a land with enough wood to build a fire.

November 2nd, 1805 [Biddle] Near the river we see cottonwood, sweet-willow, a species of maple, broad-leaved ash, purple haw, a small species of cherry, purple curant, gooseberry, red-willow, vining and whiteberry honeysuckle, huckleberry, sacommis, two kinds of mountain holly, and the common ash. The mountains on each side are covered with pine, spruce-pine, cottonwood, a species of ash, and some alder. After being so long accustomed to the dreary nakedness of the country above, the change is as grateful to the eye as it is useful in supplying us with fuel.

Lewis and Clark might not have been quite so pleased if they had realized that the change was due primarily to the increase in rainfall. Five feet of rain fall in the Columbia Gorge in a typical year. The Corps of Discovery hardly saw a dry day in the next six months.

The natives the party encountered became more and more diverse. Captain Clark became enamored with the way they made themselves beautiful.

November 1st, 1805 [Clark] Their noses are all Pierced, and they wear a white shell maney of which are 2 Inches long pushed thro' the nose. all the women have flat heads pressed to almost a point at top. They press the female childres heads between 2 bords when young untill they form the skul as they wish which is generally verry flat. This amongst those people is considered as a great mark of buty, and is practised in all the tribes we have passed on this river more or less. Men take more of the drugery off the women than is common with Indians.

Good highways follow both banks of the Columbia Gorge. Interstate 84, the least interesting, keeps close to the river on the Oregon side. On the opposite bank the State of Washington is justly proud of its Lewis and Clark Highway. The country is so beautiful it's worthwhile to drive both sides. If you do, go *upstream* on the Washington side, *downstream* on the Oregon. You'll want to tarry a lot, and the few turnouts of Washington's Lewis and Clark Highway are on the river, or right side of the road, going east.

Cascades Locks

Oregon's Cascade Locks is more or less the tourist center of the Columbia Gorge. The name comes from the navigation locks built here in

Woman and Child, Showing How the Heads of Children are Flattened: George Catlin. Lewis and Clark observed this ritual while on the lower Columbia River. The Flathead Indians did not practice this technique—their name came from the sign language used to identify the tribe. Courtesy of the National Museum of American Art, Smithsonian Institution, gift of Mrs. Joseph Harrison, Jr.

1896. The closing of the Bonneville gates took away some of the beauty, because the lovely cascade and the locks themselves were inundated. Legend has it that there used to be a natural bridge here, "the bridge of the Gods." It seems that two brothers, Klickitat (Mt. Adams) and Wyeast (Mt. Hood), were competing for the affections of Squaw Mountain. They argued, growled, and rumbled at each other, stomped their feet, spat ashes back and forth, and belched forth great clouds of black smoke. The argument grew so heated that the earth shook and the bridge collapsed into the river, taking with it Loo-Wit, the wise old woman who guarded the span. The great spirit felt badly about the faithful Loo-Wit, so he took her, made her beautiful, and settled her in the west at a place we now call Mt. St. Helens. The lovely steel replacement bridge you see today, made by man, was erected in 1926.

Cascade Falls, which Captain Clark called the "Great Shute," are now beneath the waters backed up by Bonneville Dam. Even so, the river is so narrow here and the water moves so fast that you can get an inkling of what the Corps of Discovery was up against.

October 31st, 1805 [Clark] This Great Shute or falls is about ½ a mile, with the water of this great river compressed within the space of 150 paces in which there is great numbers of both large and Small rocks, water passing with great velocity foaming & boiling in a most horrible manner

There would be no running this rapid. The next day Captain Clark "set all hands packing the loading over the portage." During the summer, Amtrak's Pioneer makes a brief stop at Cascade Locks. Those coming from Portland have enough time between the west- and eastbound trains to tour the town's two museums and have a leisurely lunch before heading home. The old locktender's house is now a historical museum. Nearby is Oregon's first locomotive, the *Oregon Pony,* which used to work the portage railway. For 35 years, until the building of the locks, this railway provided the linkage between the riverboats that plied the upstream waters and the oceangoing boats that, given a favorable tide, could sail as far up as Cascade Locks.

For years the town has been the port for Columbia Gorge cruise boats. A new addition to the fleet is the *Columbia,* a 165-foot replica of an old stern-wheeler. This three-deck beauty has a bar and a place to buy snacks. She makes three or four 2-hour trips each day during the summer. On Saturday nights she makes an enchanting dinner cruise.

Eagle Creek

The beaver is the official Oregon animal and is the name of the state college's athletic teams. Ask a Portlander where to go to see one in the wild today, and he'll probably say, "Oh, up on Mt. Hood somewhere." Much of the area surrounding the mountain is a virtual wilderness, which, surprisingly, is easily accessible to those willing to walk a bit. One of the oldest (1900) and most popular of the many long-distance trails in the Cascades follows Eagle Creek, which dumps its waters into the Columbia just above Bonneville. While there is no guarantee you'll see a beaver in the wild, if you walk the trail, you will get a tremendous sense of what this country looked like in the days of Lewis and Clark.

The trailhead is about half a mile south of the freeway on a paved road that goes past the fish hatchery. Without backpacking equipment, it isn't possible to walk the entire trail, but consider going partway at least. There are 7 waterfalls along the 13-mile trek to Wahtum Lake. The lower section is close to the creek with its great salmon spawning grounds. The first fall, Metlako, is only a 1½-mile walk. Another half mile brings you to the Punchbowl Falls, one of the loveliest in the canyon. The first campground is 3½ miles from the trailhead, just above High Bridge. Tunnel Falls, so named because the trail leads behind the falls, is near milepost six. Though this trail becomes crowded at times, particularly on weekends, it is still a marvelous way to see an unspoiled part of the Cascades.

Bonneville Dam

Oldest of all the Columbia dams, Bonneville has been a symbol of progress since the time it was built during the Great Depression. The dam was named for the nearby stop on the Union Pacific Railroad, which in turn was named after an early explorer, Captain Benjamin Louis Eulalie Bonneville. Bonneville's career is shrouded in mystery. Born in France, he attended West Point and was given a commission in the U.S. Army. In 1831 he was granted a two-year leave of absence, obstensibly to allow him to pursue the fur trade. He never seems to have trapped much fur, and he overstayed his leave by two years, leading to speculation that he was actually a spy, sent west by Andrew Jackson to see what the British were up to. His most notable achievement was to send his lieutenant, Joseph Walker, to California, where he more or less charted the course the forty-niners would later use to get to the gold fields.

Bonneville made two trips to Oregon and returned lauding its virtues. Describing the Willamette Valley, he wrote: "One of the most beautiful, fertile and extensive vallies in the world, wheat, corn and tobacco country. If our Government ever intend taking possession of Origon the sooner it shall be done the better." Spy or no, he was regranted his commission, but thereafter always found himself assigned to what the army considered the worst posts—Fort Vancouver, for example. His fame is due primarily to Washington Irving's accounts of his adventures.

Bonneville Dam was begun in 1933, and the gates were closed five years later. The cost of the dam, navigation locks, and powerhouse was $88 million. By contrast, the recently opened generating facility alone, on the Washington side, cost more than $600 million. This and all other dams on the Columbia were built by the Army Corps of Engineers. The electricity is sold to a public agency, the Bonneville Power Administration, which distributes it to private utilities throughout the Northwest. It's a good deal for the rate payers: The electricity is sold for about a quarter the price of nonsubsidized power.

The Corps of Engineers has come in for a great deal of criticism in some quarters, particularly from those who blame the dams for the drastic drop in the salmon catch. For this reason, the Corps spends a lot of money getting its side of the story across to those who visit the facilities. The visitors center at Bonneville has extravagant displays, and all along the river, from Bonneville to Idaho, the Corps has built countless facilities to make it easy for the public to enjoy the recreational possibilities of the various projects. Whatever their motives are, the visitors centers are fun to visit. The one at Bonneville has a lovely historical and Native American exhibit as well as an instructive presentation on how power is generated.

The star attraction is the observation room for the fish ladder. Here you can actually see the migrating salmon, coho, and steelhead as they are being counted by a naturalist in an adjacent room. The Corps has spent millions of dollars on hatcheries here at Bonneville and at many more upstream. Driving along the river you will see gleaming stainless steel tank-trucks, outfitted with the most expensive gauges and pumps, which are hauling fingerlings downstream so they won't get caught in the turbines. Unlike those at Bonneville, recently built fish ladders have been routed away from the turbines and gates so the migrating fish don't find themselves swept right back down again. Has all this effort been enough to compensate for the loss of spawning grounds? The jury is still out, though there is little doubt that ecologically the dams have been a mixed blessing.

The navigation lock is quite interesting. With a lift of 70 feet, it's not the highest on the river, but it is one of the most visible. Navigation is free to all users, so to that extent the government is subsidizing all those barge operators you see steaming by.

The Columbia Gorge Falls

Leave the interstate just west of Bonneville Dam and take the old road, which leads past seven spectacular waterfalls, more than you'll find in any other comparable place in the United States. Geologically, these are young mountains; there hasn't been the time for the streams to wear them down, so the canyons are V-shaped. The basaltic nature of the rock creates hard spots that resist erosion. This, together with lots of rain, provides all that is needed to give the gorge its majestic cataracts. On the downward trip the explorers didn't have much to say about the falls, but on their return they were in less of a hurry, and so had time to look around.

April 9th, 1806 [Lewis] we passed several beautifull cascades which fell from a great hight over the stupendious rocks which closes the river on both sides. the most remarkable of these cascades falls about 300 feet perpendicularly over a solid rock into a narrow bottom of the river on the south side. several small streams fall from a much greater hight, and in their decent become a perfect mist which collecting on the rocks below again become visible and decend a second time in the same manner before they reach the base of the rocks.

Each of the waterfalls is lovely in its own way. Many can be seen without leaving your car, but don't try, because to appreciate them you must use all your senses. Waterfalls have a living as well as a scenic quality; your eyes see the beauty, your ears sense the power, your skin feels the freshness, and your nose provides the piece de resistance, the sweet smell of the pine forest. Here, listed from east to west, are the principal falls on the Oregon side of the gorge:

Tanner Creek. The trailhead is directly opposite the Bonneville exit on I-84. The 1-mile path to the falls is narrow and subject to slides.

McCord Creek. Park your car at the Yeon State Park near the hamlet of Warrendale. The lower falls, less than a half mile from the road, are easily accessible by a wide path. A steep trail then goes up the left bank of the canyon a mile and a half to the upper falls. At one point the trail runs along an overhanging cliff and is quite narrow. An iron railing protects you from the steep drop.

Horsetail Falls/Oneonta Falls. Take the old "Scenic Highway" to a point half a mile west of Ainsworth State Park. Horsetail Falls are so close to the parking lot you often have to keep your windshield wipers on to see through the mist. One of the most popular trails in the gorge goes up the right bank to the upper falls, then contours around the hill to Oneonta Creek, where a short trail leads up to the lovely Triple Falls. A trail down the canyon leads to the old road from which it's a short walk back to your car.

Multnomah Falls. These are the most famous of all of Columbia's wonderful cataracts. They are either the third highest or the fourth highest falls in the U.S., depending on what criterion one uses to define height. Hundreds of thousands of tourists a year walk the easy paved loop trail to the bridge between the upper and lower falls, and then return to the restaurant and gift shop across from the parking lot. For a different view take the wide but steep trail to the top of the upper falls. You can then either return on the same trail or walk around the hill to the next canyon and descend to the Wahkeena Falls. From there a trail paralleling the highway leads back to your car.

Wahkeena Falls. The name is Indian, meaning "most beautiful," and you may agree that they are. Park at the campground and take the short trail to the bridge spanning the river just below the falls. A bit to the west of the parking lot is another cascade, Mist Falls, one of the more delicate in the area.

Latourell Falls. Some people, this author included, believe this pair of falls is the most lovely in the gorge. The lower falls are close to the parking lot. A trail goes down to the pool at the base, where you can bathe in the wispy vapors thrown off by the wind. A well-marked trail climbs up the left side of the canyon, where you get a constantly changing view of the lichen-covered basaltic cliff and the boisterous torrent of water. You're almost never out of earshot of its roar. At the lip of the lower falls, the trail makes a fork. Take the left branch up the canyon another half mile to the upper fall. You gaze in awe at a 15-foot-wide ribbon of shimmering white foam, which comes plunging over a cliff, dropping a hundred feet to a point where it is suddenly scooped up by a plow-shaped outcropping of basalt and funneled into a teapot spout, which in turn is thrown off to the left where it drops another 50 feet before being dashed to smithereens in the pool below. If you can tear yourself away from this enchanting spot, cross the little bridge at the base of the falls and take the right-hand trail back to the fork.

Crown Point. West of Latourell Falls, the road climbs steeply in some more of those lovely loops that make this highway so poetic. At the top you circle the Crown Point vista house. This impressive structure, built in 1917, occupies what certainly is the most photographed area of the gorge. The view is spectacular. The Portland Women's Club built another viewpoint a mile and a half farther west.

Larch Mountain

If you find yourself touring the gorge on a clear day, by all means take the side road to the top of Larch Mountain. The paved road to the summit joins the Scenic Highway about a half mile west of Crown Point. This 10-mile-long road climbs 4000 feet to a belvedere overlooking the most scenic volcanos in the Cascades. To the south loom Mt. Hood and Mt. Jefferson. Across the river, Mt. St. Helens is on the left, Mt. Adams on the right, and in between is the giant of them all, the 14,410-foot Mt. Rainier, its glaciers shimmering in the sun.

The Oregon side of the gorge ends at the Sandy River campground. A short nature trail describes some of the native plants in the area. Apparently, the Corps of Discovery's diet was improving.

November 3rd, 1805 [Clark] Capt. L. and 3 men set out in serch of the Swans, Brants Ducks &c. &c. which appeared in great numbers. he killed a Swan and Several Ducks, which made our number of fowls this evening 3 Swan, 8 brant and 5 Ducks, on which we made a Sumpteous supper.

It was here, on their return trip, that Lewis and Clark realized they had missed the entire Willamette Valley, so Clark backtracked to take a look. We'll take a look, too, but first, let's also backtrack to Cascade Locks and explore the Washington side of the river.

Stevenson

Just east of the Bridge of the Gods is the site of Fort Cascades, scene of a three-day battle in 1856. A band of Yakimas and Klickitats, led by Chief Kamiakin, caught a detachment of soldiers and settlers holed up in the fort and tried to drive them out. About this time the riverboat *Mary* arrived with enough guns and ammunition to keep the Indians at bay. The boat escaped and got the word to Fort Vancouver, where Lieutenant Phil Sheridan was stationed. This colorful officer, whose dashing escapades during the Civil War made him a national hero, arrived with

40 dragoons who put the marauders to rout. It was the last opposition to white settlement on the lower Columbia.

Beacon Rock

One of the most remarkable spots in the Columbia Gorge is located on the Washington side, about 8 miles downstream from the Bridge of the Gods.

October 31st, 1805 [Clark] A cloudy rainey disagreeable morning I proceeded down the river to view with more attention the rapids we had to pass on the river below. A remarkable high detached rock Stands in a bottom on the Stard. Side near about 800 feet high and 400 paces around, we call the Beacon rock.

Captain Clark was a remarkable observer; the U.S. Geological Survey gives the height of the rock at 850 feet, or just a tad more than 800 feet above the level of the river. Not only is Beacon Rock a landmark, it also provides the best belvedere for viewing the entire Columbia Gorge. At first glance it seems obvious that ropes, pitons, carabiners, and the rest of the rock climbers' paraphernalia will be required, but thanks to a Portlander, Henry J. Biddle, you can now climb to the dizzy summit with some ease. Biddle bought the rock with the express purpose of preserving it and building a trail to the summit. In 1924 he wrote an article for the Oregon Historical Society:

Work was commenced on the trail in October, 1915, and it was completed in April, 1918 . . . The trail is about 4,500 feet long, 4 feet wide, and with a maximum grade of 15 percent. It extends from the North Bank Highway . . . to within about 20 feet of the summit. The rock there becomes so narrow that the construction of a wide trail was impracticable, so a narrow flight of steps leads to the topmost point. There are 52 hairpin turns in the trail, 22 wooden bridges, and over a hundred concrete slabs, spanning the minor fissures in the cliff.

The State of Washington now maintains the trail. Those not afflicted with vertigo will not only find the view spectacular, but will marvel at this engineering accomplishment. In several places the cliff is so steep, you cross under a bridge that soon proves to be your trail—it has turned backward on itself and has begun going the other way. A steel-pipe handrail is the only thing between you and thin air.

The same day the Corps of Discovery passed Beacon Rock, Captain Clark noted that they had reached tidewater, though he was quick to surmise—correctly, as usual—that during the high spring floods the tides would no doubt commence much farther downstream. Many of the party had been complaining about the fleas they encountered at the lodges of the local tribes. Now Clark was beginning to get angry about thievery:

November 4th, 1805 [Clark] dureing the time we were at dinner those fellows Stold my pipe Tomahawk which they were Smoking with, I imediately serched every man and the canoes, but could find nothing. while Serching for the Tomahawk one of those Scoundals Stole a coat of one of our interperters, which was found Stufed under the root of a tree, we became much displeased with those fellows, which they discovered and moved off.

Mt. St. Helens

That same day, Captain Clark also wrote:

November 4th, 1805 [Clark] we had a full view of Mt. Helien which is perhaps the highest pinical in America [from their base] This is the mountain I saw on the 19th of October last covered with Snow, it rises Something in the form of a Sugar lofe.

Many of the sights we have seen on this odyssey have changed since the days of Lewis and Clark, but none more spectacularly than Mt. St. Helens. It doesn't look like a sugar-loaf anymore. What the explorers saw was not the "highest pinical in America," but nevertheless a mountain of spectacular beauty that rose 9667 feet into the sky. On May 18, 1980, the upper 1300 feet were suddenly gone, thrown 10 miles into the air by the biggest eruption scientists have ever been able to see and photograph first-hand. It wasn't a first for Mt. St. Helens, though. John C. Fremont, while camped near The Dalles in 1843, reported an eruption that "lit the night skies." The Canadian artist Paul Kane made a painting of an eruption he observed a few years later, which now hangs in Toronto's Royal Ontario Museum.

The site of the eruption is now called the **Mt. St. Helens National Volcanic Monument,** administered by the Forest Service. It takes a full day to really see the destruction, regardless of whether you start here on the Columbia or farther north, along Washington's Interstate 5. The best way to see it is to start at the hamlet of ***Carson,*** at the mouth of the

Wind River. A lovely paved road wends its way through thick stands of Douglas fir as it climbs to Outlaw Ridge, where you get your first sight of the mountain itself. The road then drops into the Lewis River Valley, named, of course, after our hero. The staff at a Forest Service Information Center here can advise you on whether or not you can go any farther. The Forest Service, mindful that many sightseers were killed by the blast, has established red and blue zones that may be closed if things start acting up again. Most of the roads into the area are dirt, and remain unplowed until early summer. Because of the danger of fire, many are closed a month or so later.

If the roads are open, the sights are well worth the drive. You first see the havoc wreaked by mud flows along Lewis River, and then you enter an area of pristine beauty that was shielded from the blast by an intervening ridge. After climbing past Bear Meadow you are suddenly thrust into a country of unbelievable devastation. Trees, looking like matchsticks blown over by a giant, have been left where they fell. A dirt road leads to Windy Ridge, barely 4½ miles from the crater's center. Everything seems dead, but if you look carefully under the pumice you'll see nature starting to take hold again. Park rangers give intrepretive lectures at Windy Ridge. You can return to the Lewis and Clark trek by driving to Portland via the Lewis River road or continuing north to the town of **Randle,** where you pick up U.S. Highway 12, which goes west to Interstate 5. The Forest Service operates a large visitors center on a side road near the Interstate.

THE PACIFIC COAST RANGE

We left the Corps of Discovery on the western edge of the Columbia Gorge, camped at the mouth of the Sandy River, which they called the Quicksand. On their westward journey the explorers didn't find what is, by all accounts, the greatest natural resource in the Pacific Northwest, the fantastically fertile Willamette Valley. Their initial failure is understandable: They were in a hurry, they had found tidewater, and they knew the long sought Pacific was near. Furthermore, below the gorge the river becomes wide and slow-moving, with numerous islands hiding myriad channels. They missed the mouth of the river on their return trip, too, but they were much too good explorers not to realize something was amiss.

April 1st, 1806 [Lewis] the Indians who encamped near us last evening informed us that the quicksand river which we have heretofore deemed so considerable, only extendes through the Western mountains as far as the S. Western side of mount hood. we were now convinced that there must be some other considerable river which flowed into the columbia on it's south side below us which we have not yet seen.

April 2nd, 1806 [Lewis] several canoes of the natives arrived at our camp. these men informed us that 2 young men whom they pointed out resided at the falls of a large river which discharges itself into the Columbia on it's South side some miles below us. we redily prevailed on them to give us a sketch of this river which they drew on a mat with a coal. it appeared that this river which they called Mult-no-mah discharged itself behind the Island which we called the image canoe Island; they informed us that it was a large river and run a considerable distance to the South between the mountains. Capt. Clark determined to return and examine this river

April 3rd–4th, 1806 [Biddle] 13 miles below the last village, he entered the mouth of a large river, which is concealed by three small islands. Cpt. Clark now discovered to the southeast a mountain which he had not yet seen, to which he gave the name of Mount Jefferson. Like Mount St. Helen's, its figure is a regular cone, covered with snow; The current of this river is as gentle as that of the Columbia; its surface is smooth and even, and it appears to possess water enough for the largest ship. It's length from north to south we are unable to determine, but we believe that the valley must extend to a great distance. It is in fact the only desirable situation for a settlement on the western side of the Rocky mountains, and, being naturally fertile, would, if properly cultivated, afford subsistance for 40,000 or 50,000 souls.

Captain Clark had not only found the shangri-la of the Pacific Northwest, but had the insight to recognize its potential. In one fell swoop he became the first white to set eyes upon the future site of Portland, and he found the entrance to the valley that would become the magnet for willing pioneers to brave the rigors of the Oregon Trail. The weather must have been good that day; Mt. Jefferson lies 90 miles to the southeast and most days it's hidden by clouds.

Multnomah became the name of Portland's county, but the river and its valley are called the Willamette, pronounced Will-AM-it. It wasn't long before people were lauding its virtues. In 1836 Washington Irving wrote in *Astoria:*

Through the centre of this valley flowed a large and beautiful stream, called the Wallamot, which came wandering, for several hundred miles, through a yet unexplored wilderness. The sheltered situation of this immense valley had an obvious effect upon the climate. It was a region of great beauty and luxuriance, with lakes and pools, and green meadows shaded by noble groves.

Gregory M. Franzwa, in his book *The Oregon Trail Revisited,* put it in more blunt terms:

There was a valley out there; a long, wide one where the ground was more fertile than anywhere else on earth. There was plenty of rainfall, it was warm in the winter, comfortable in the summer, and the Indians couldn't care less if any white man came around—there was plenty of land along the Willamette.

By 1850, 13,000 people had settled here; it had become the Oregon Trail's raison d'etre. Historians are unsure just how far up the river Captain Clark went; probably not quite to what is now downtown Portland. He most certainly didn't go another 12 miles upstream, for he would then have seen and commented on the tremendous falls at Oregon City. Today, more than 60% of Oregon's population resides in the Willamette Valley. Between Portland and Eugene, Oregon's second largest city, lies the state capital, Salem, the college town of Corvallis, Albany, and dozens of smaller farm villages.

Portland

The flip of a coin determined that you are now in Portland, not Boston, Oregon. The two earliest settlers each wanted to name it after their New England hometown. Over 1.3 million people live in "The City of the Roses" and its environs, making this the first real metropolitan area we have visited since leaving Omaha. The city straddles the Willamette, with most of the commercial center lying on the east bank, where a grassy esplanade follows the river for nearly a mile. Architecturally, downtown Portland is as exciting as any city in America. Old cast-iron light standards grace the sidewalks, yet many of the new buildings are bold in design, with pleasant courtyards sporting fountains and statues. Of particular interest is the spanking new Pioneer Courthouse Square. The financing scheme for the plaza was an act of genius. Realizing that Portlanders didn't want their downtown to become a wasteland after business hours, they hit on a plan to sell bricks. Fifty thousand people paid $5 each to have their name stamped on a brick. Those bricks now pave a city block, and it's heartening to see locals, flashlight in hand, looking for the brick with their name on it, even late into the evening. Not only did the financing scheme raise a lot of money, but the contributors now feel they have a personal stake in their community and its downtown. Free entertainment is often provided, and the police are careful to keep any single group of people from laying claim to the area. In a further effort to keep its downtown vital, Portland has embraced mass transit in a big way. The main shopping streets have been made into pedestrian and bus malls, and a new subway is just coming into use.

Portland is a great walking town. Stop in at the **Convention & Visitors Association** on lower Salmon Street and pick up a copy of the brochure *Portland Walk, Ride, Drive Guide*. Then head out. The old area, south of the Amtrak station and toward the river, has been renovated with "antique" stores, trendy bars, boutiques, and art galleries.

The most recent renovation is an old brick warehouse, called **New Market,** which has been transformed into a restaurant and shopping mall as exciting as Boston's Faneuil Hall or San Francisco's Ghirardelli Square. Neo-Greek columns, one collapsed in a rather whimsical fashion, stand in front of the plaza. An outdoor market, somewhat in the European tradition, is held on Saturdays and Sundays near the waterfront beneath the Burnside Bridge.

Old-town Portland streets are still lit by 100-year-old cast-iron lamp posts, and the sidewalks are graced by a number of bronze, four-spout drinking fountains, a gift of the lumber baron Simon Benson. He hoped the fountains would tempt his thirsty lumberjacks away from the numerous saloons of the day. One of Benson's more workable projects was the establishment of a first-class hotel. For 50 years the **Westin Benson** has been the grand dame of all Pacific Northwest hotels, though there are now several other downtown hotels that are just as nice. Those who are not on a first-class budget but still want to stay downtown might consider staying at the **Imperial Hotel.**

Portland has a fine public art gallery on Park Street. Nearby is the **Oregon Historical Society,** where you can see exhibits showing how the tribes of the lower Columbia lived before the arrival of the white man. The Georgia Pacific lumber company has an historical museum of early lumbering practices in the basement of its skyscraper. The city is home to **Lewis and Clark College, Portland State University,** and **Reed College,** a small liberal arts school with a national reputation. The famous **Rose Garden** is uphill to the west of downtown, while the equally beautiful **Rhododendron Garden** is in the southeast, across from Reed College. **Washington Park** has a first-class zoo and a fine statue of Sacagawea sculpted in 1904 by Alice Cooper to commemorate the centennial of the Lewis and Clark expedition.

River Cruises Not surprisingly, Portland is the principal port of the Columbia River. Goods brought here by barge from the Inland Empire are loaded onto ships destined for ports throughout the world. It's a pity, but Alaska-bound cruise ships usually don't make this vibrant city a port of call. There is, however, one cruise line operating out of Portland. During the summer an Exploration Cruise Lines ship makes week-long voyages through the country we have been exploring. About 100 passengers are carried aboard a modern vessel that first sails downstream to call at Astoria for a trip to **Fort Clatsop.** The ship then turns back upstream, sailing all the way to Lewiston, Idaho. Cruise passengers have time to see many of the sites we have been seeing: Bonneville Locks and Dam, Maryhill Museum, the Whitman Mission, and the Nez Perce

Museum at Spalding. There's even time for a jet-boat ride into Hells Canyon. It's a relaxing way to travel, and you get perhaps the best sense of what the Corps of Discovery saw, at least on the lower, undammed section of the river.

Oregon City Portland wasn't always the main event in these parts. In earlier days Oregon City had far more people, drawn here because of the site's proximity to the great falls of the Willamette, where electricity could be generated to power lumber, grist and, later, paper mills. Oregon City boasts many firsts in the Pacific Northwest: first territorial capital, first Catholic Archdiocese, first Masonic Lodge, first navigation locks, and the first electrical power transmitted over any great distance. The mills were built in 1866; the navigation locks around the falls were finished in 1873.

At first glance Oregon City is nothing more than a drab paper-mill town. The ugly Weyerhauser plant is on one side of the falls, Crown Zellerbach's on the other. The townsfolk, though, remembering their important role in the development of the Pacific Northwest, have preserved some interesting buildings and have built a promenade along the cliff bordering the falls. Oregon City is really two towns, one down by the river, the other up on a bluff. To get from one to the other, you ride the municipal elevator. The new concrete structure, with its enclosed observation room at the top, replaced an old free-standing corrugated iron tower that was reached by crossing over the railroad tracks on a rickety bridge several hundred feet above the ground. The elevator leads to the promenade and to the nearby **McLaughlin House,** which is now a museum. John McLaughlin, "the father of Oregon," retired to Oregon City and became an American citizen after he left the Hudson's Bay Company. His 1845 Georgian-style frame house contains most of his possessions, and in some ways you learn more about the fur trade here than at the National Park Service museum in Vancouver, Washington, where he was the chief factor for many years. A statue of McLaughlin graces the halls of the Capitol in Washington, DC. Oregon City is the county seat for Clackamas County, whose historical society has exhibits of period rooms, a working kitchen, and an Oregon Trail interpretative center. Captain Clark, incidentally, spelled the name of the tribe for which the county was named *Clark-a-mus.*

Vancouver

Vancouver, Washington, was the site of an Indian village when the Corps of Discovery camped there on its downstream voyage. The influence of the sea-coast traders, particularly the British, was already being felt this far upstream.

November 4th, 1805 [Clark] The Indians which we have passd today were of the Scil-loot nation differ a little in their language from those near & about the long narrows. their dress differ but little, except they have more of the articles precured from the white traders they all have flatened heads both men and women, live principally on fish and Wap pa too roots. dureing the short time I remained in their village they brought in three Deer which they had killed with their Bow & arrows. They are thievishly inclined as we have experienced.

Although Vancouver, Washington, is a thriving, bustling city of 45,000 people, it suffers from its reputation as Portland's poor cousin, and smarts at the fact that most people confuse it with that other Vancouver in Canada. For 20 years, though, this was the most important place in the Pacific Northwest, and it was thoroughly British. Captain Gray had established an American claim to Oregon in 1792 that was reinforced by Lewis and Clark in 1805 and then by the Astorians in 1811. By the end of the War of 1812 it was the British-owned Northwest Company that had the only presence in the area. Captain Gray sailed home, Lewis and Clark went back to Missouri, and when the British Navy stood off Fort Astoria in the frigate *Beaver* John Jacob Astor's men knew when to back down. Several of the Astoria partners were formerly Northwest Company men, and they simply sold out and went home, too. The Treaty of Ghent, which formally ended the War of 1812, defined the United States–Canadian boundary east of the Rockies at the 49th parallel, but called for joint ownership of Oregon. The British, no doubt, expected the line to eventually be drawn at the Columbia River. Russia had some claims north of 54°40′, but they never attempted to enforce them. No one consulted the Chinooks, Clatsops, Klickitats, Cowlitz, Umatillas, or Wallawallas. Since there were no Americans in Oregon in 1814 the only challenge to the Northwest Company's dominance came from a British rival, a firm formally known as the Gouvenors and Company of Adventurers of England Trading into Hudson Bay. Those battles were being waged in Canada, but the final outcome was to have an effect on the Oregon country. In 1821, tired of slugging it out, the two firms merged, with the Hudson's Bay Company the survivor.

Dr. John McLaughlin was sent out to manage things. This 6-foot-4-inch Scotsman, who sported "a beard that would do honor to the chin of a grizzly bear," had exactly the right talents for taming the Pacific Northwest. His first step was to close forts Astoria, Walla Walla, and Spokane, and consolidate operations at Fort Vancouver, which he built in 1824. He sent trapping brigades as far inland as the Great Salt Lake and planted nearby fields in peas, oats, barley, wheat, and garden vegetables. He established orchards of apples, pears, peaches, plums, and cherries, and kept herds of cattle, horses, hogs, sheep, and goats. The fort echoed to the sounds of carpenters hammering and sawing, blacksmiths repairing tools, and coopers making barrels. It was a thriving center of English culture and authority. The fort's very success contributed to its downfall; when American settlers, weary and impoverished by the long overland trip on the Oregon Trail, needed supplies, the only place to turn was to Dr. McLaughlin's British trading post. The good doctor, openly sympathetic to the settlers, had a falling out with his superiors in Montreal and retired to Oregon City in 1846. By that time the beaver hat was out of style, the animals had been trapped to near extinction, and the Hudson's Bay Company was concentrating on other matters. The politicians, who in 1844 had shouted "54°40' or Fight" were voted down, and it was time for compromise. The treaty of 1846 extended the boundary at the 49th parallel all the way to tidewater in Puget Sound. In the end, however, it was the settlers, abetted by Dr. McLaughlin, who most affected the American case, for unlike the trappers, they had a permanent reason for being here.

For a while the fort continued to be a trading post, but it was closed in 1860. The United States Army built a new fort nearby, and the old dowager was left to rot. The National Park Service now maintains the site, and has reconstructed the walls, bastion, blacksmith's shop, bakery, and the Chief Factor's residence. A nearby museum has extensive exhibits of life at the fort and shows slides and movies depicting the events that occurred here.

Through the years, many soldiers who later would become famous officers were posted at Vancouver Barracks. The first commandant was Captain Bonneville, who hated the place, as did U. S. Grant, who was stationed here during his early military career. It was not a happy time in the life of the future president. One of his many business failures occurred when he tried to become a potato farmer, only to see his entire crop ruined by flooding of the Columbia River. Many historians attribute the start of his love affair with "Old John Barleycorn" to his loneliness in this remote outpost and his dislike of his superior officer. The house he lived in, the oldest on the post, is now a museum. The **Clark County Museum** is housed in a 1909 library and includes a pioneer doctor's office, country store, Indian artifacts, and a railroad exhibit. The ornate Victorian-style 1885 home of L. M. Hidden now houses a fine restaurant and art gallery.

Twenty miles north of Vancouver, on road 503, is the little town of **Battle Ground**, home of the Chelatchie Prairie Railroad. During the summer you can ride behind an old steam logging locomotive as it makes a 10-mile journey to Yacolt. The route goes through lovely timber country, dives into a 300-foot tunnel carved out of solid rock, and then crosses an 85-foot-high trestle that spans the Lewis River.

Sauvie Island

The trip westward from the Portland/Vancouver area can be made on either the Washington or Oregon side of the river. The most interesting way is to combine a little of both. Start downstream from Portland on U.S. Highway 30. The first 10 miles or so are through a rather bleak industrial area, but even this is preferable to the interstate, which follows the river for 40 miles on the Washington side. Soon after passing the beautifully ornate St. Johns Bridge, the countryside becomes rural. To get a hint of what Lewis and Clark saw, turn off the highway, cross a slough, and make a short detour to Sauvie Island.

November 4th, 1805 [Clark] This Island is 6 miles long and near 3 miles wide thinly timbered. we landed at a village of 25 houses: 24 of those were thached with straw, and covered with bark. I counted 52 canoes on the bank in front of this village maney of them verry large and raised in bow. we recognized the man who over took us last night, he invited us to a lodge and gave us a roundish roots about the Size of a Small Irish potato which they roasted in the embers until they became Soft, This root they call Wap-pa-to.

On its return trip, the Corps of Discovery named this piece of land Wappatoo Island; the explorers had become fond of the root that nowadays people call "swamp potato." A Bostonian merchant, Nathaniel Wyeth, set up a trading post on the island in 1834, which he hoped would be the Pacific Northwestern version of his very successful Fort Hall, located near present-day Pocatello, Idaho. It was not to be; the post lasted only one season. The Oregon Historical Society has refurbished the old Bybee-Howell farmhouse as a museum. Turn left at the fork and drive about a mile west on the levy road. Along the way you'll see a common sight along the rivers of Washington and Oregon—millions and millions of logs tied into rafts and secured to the shore. These sloughs and backwaters are the warehouses of the lumber industry.

Cathlamet

Just before reaching the paper mill at Westport on Highway 30, turn right to take the only remaining ferry on the Columbia River. The little diesel-powered, nine-car vessel makes hourly crossings from early morning to late at night. If you just missed a boat, head back toward the highway and have a beer in Westport. The back-bar in the town's only saloon is a beauty, brought around the Horn on a sailing ship. The ferry ride is regretfully too short, about 15 minutes, but you do get a first-hand view of the river as the pilot deftly maneuvers the craft around fishing boats, floating logs, and oceangoing ships. After crossing an island you come to the town of Cathlamet, pronounced Cath-LAM-et, Washington, one of the prettiest on the river. The town was named for the tribe the Corps of Discovery met here on November 6th. Cathlamet has an interesting bar, too, where you can wile away the time waiting for the ferry. It also has an unpretentious museum, chockfull of all those things people hate to throw away because they're old but don't know quite what else to do with. Old dresses are in one corner, milking machines in another. The hodgepodge is delightful.

At Crown Zellerbach's lumber mill just west of town you'll see an old narrow-gauge locomotive that used to haul logs down from the mountains. This old Shay, as she's called, has three pistons, located vertically alongside the boiler. The rods drive gears, which in turn drive the tiny wheels. Shays couldn't go very fast, but they could pull a mighty load. There are a number of places in the Pacific Northwest where you can learn about old-time logging practices. One, located on U.S. Highway 26, is about 20 miles west of where Lewis and Clark established their salt cairn at Seaside. The most extensive logging museum, however, is on U.S. 97, about halfway between Bend and Klamath Falls, Oregon. Here, out in a yellow pine forest, stands an amazing collection of locomotives, steam tractors, donkey engines, and power saws, all looking like they're just waiting for someone to come along, light their boilers, and put them to work.

The road west from Cathlamet passes through grazing bottomland and hillsides partially covered with forest. This is clear-cutting country, where whole tracts are denuded of every tree larger than a sapling. If loggers are working near the highway, you can see the techniques employed. Often one tree is left standing to serve as a mast for the high-wire. "Chokers" roam the steep slopes, snaking cables around the logs in what is probably the most dangerous operation in the woods. Crawler tractors, with iron jaws as big as an automobile, gobble up logs 6 feet in diameter and pile them onto 18-wheel trailer rigs. Here you get a sense of the life that author Ken Kesey described in his novel *Sometimes a Great Notion.* Is clear-cutting the best way to husband our resources? Forest products industry economists say yes, but a stroll through a previously clear-cut area reveals a new growth of junk plants, alder, dogwood, and Scotch broom, all of which grow like weeds in this damp climate and seem to crowd out the better species.

A violent storm erupted just before the Corps of Discovery was to begin the most dramatic portion of its trip, yet Captain Clark was still intent on describing in minute detail what he saw. Perhaps anticipating the Victorian Age, he commented on the dress of the local women.

November 7th, 1805 [Clark] The garmet which ocupies the Waist and thence as low as the knee cannot properly be called a petticoat, in the common acceptation of the word it is a Tissue formed of white cedar bark which are interwoven in their center by means of Several cords which Serves as well for a girdle and which Strans, confined in the middle, hang with their ends pendulous from the waist, the whole being of Sufficent thickness when the female Stands erect to conceal those parts useally covered from familiar view, but when she stoops or places herself in any other attitude this battery of Venus is not altogether impervious to the penetrating eye of the amorite.

Altoona

As you drive west on State Highway 4, take time to detour a half mile to Washington's only remaining covered bridge near the hamlet of Skamokawa. Built in 1905, it spans a stream meandering through a lush meadow. Holstein cows graze near curved-roof barns, giving this country an Old-World, Dutchlike flavor.

At the town of **Rosburg,** make another detour 7 miles south to Altoona, an almost deserted village that is a shrine for Lewis and Clark buffs. There are no merchants in Altoona anymore. The salmon cannery closed years ago. The rotting building, teetering atop barnacle-covered pilings, looks as if it will soon collapse into the river. Then there will be nothing in Altoona save a half dozen beautiful homes, perched on the steep wooded bluff overlooking an impressive rock jutting out of the water.

November 7th, 1805 [Clark] A cloudy foggey morning Some rain. we Set out early proceeded under the Stard. Side under a high rugid hills. We proceeded on about 12 miles below the Village under a high mountaneous Countrey boald and rockey and Encamped under a high hill on the Stard. Side opposit to a rock Situated half a mile from the shore, about 50 feet high and 20 feet Deamieter;

The fog forced the Corps of Discovery to land near what is now called Pillar Rock. Then, near midday, it cleared, and the great captain, who seldom betrays a feeling or indulges in sentiment, penned in the logbook that he kept fastened to his knee: "Ocian in view! O! the joy." His journal for that day reads:

November 7th, 1805 [Clark] Great joy in camp we are in view of the Ocian, this great Pacific Octean which we been so long anxious to See. and the roreing or noise made by the waves brakeing on the rockey Shores (as I suppose) may be heard distinctly.

The ocean cannot possibly be seen from this point, but, given the intensity of the weather, it's not hard to understand why Captain Clark considered the wind-blown waves to be of ocean-size proportions. They still had another 15 miles to go and it would be another week before the Corps of Discovery finally set eyes on its objective. Nonetheless, they kept inching their way along the north shore. To follow their route, return to Highway 4, drive west to **Naselle,** and then go south on Highway 401 to the hamlet of **Megler.**

November 8th, 1805 [Clark] after Dinner we took the advantage of a returning tide and proceeded on to the Second point on the Std. here we found the Swells or Waves so high that we thought it imprudent to proceed; we landed unloaded and drew up our Canoes. we are all wet and disagreeable, as we have been for Several days past, and our present Situation a verry disagreeable one in as much, as we have not leavel land Sufficient for an encampment and for our baggage to lie cleare of the tide, the High hills jutting in so close and steep that we cannot retreat back, and the water of the river too Salt to be used. added to this the waves are increasing to Such a hight that we cannot move from this place, The Seas roled and tossed the Canoes in such a manner this evening that Several of our party were Sea Sick.

The party deserved better shrift, but the storm didn't give an inch.

November 9th, 1805 [Clark] the tide which rose untill 2 oClock P M to day brought with it such emence swells or waves, aded to a hard wind from the south which loosened the drift trees which is verry thick on the shore, and tossed them about in such a manner, as to endanger our canoes very much. notwithstanding the disagreeable Situation of our party all wet and cold they are chearfull and anxious to See further into the Ocian.

Point Ellice

The next day the wind abated, and they were able to move camp to the lee of Pt. Ellice, where they camped for four nights. Point Ellice is the northern anchorage for the bridge that now spans the Columbia. A sign along the highway, a half-mile east of the bridge, marks the place Captain Clark called Camp Distress.

November 12th, 1805 [Clark] A Tremendious wind from the S.W. about 3 oClock this morning with Lightineng and hard claps of Thunder, and Hail which Continued until 6 oClock a.m. when it became light for a Short time, then the heavens became sudenly darkened by a black cloud and rained with great violence. the waves tremendious braking with great fury against the rocks and trees on which we were encamped. our Situation is dangerous.

Not since those horrid days in the Bitterroots has the captain seemed so angry.

November 15th, 1805 [Clark] The rainey weather continued without a longer intermition than 2 hours at a time, from the 5th in the morning untill the 16th is eleven days rain, and the most disagreeable time I have experenced confined on a tempist coast wet, where I can neither git out to hunt, return to a better situation, or proceed on; in this situation have we been for Six days past.

It still rains a lot here, but one local person was overheard to say: "Yeah, but I haven't shoveled an inch of rain since I moved here from the Midwest thirty years ago." By taking advantage of a favorable tide, the explorers were able to round the point and camp just west of present-day **Fort Columbia,** a state park that has a small museum in one of the barracks. Just west of the fort, a sign marks the spot where the Corps of Discovery camped for 10 days, and where they first actually sighted the Pacific.

Nov 16th, 1805 [Biddle] Our camp is in full view of the ocean, on the bay laid down by Vancouver, which we distinguish by the name of Haley's bay, from a trader who visits the Indians here, and is a great favorite among them. It was now apparent that the sea was at all times too rough for us to proceed further down the bay by water. We therefore landed, and having chosen the best spot we could select, made our camp of boards from the old Chinook villages.

Haley was the captain of an American brig, whom Lt. Broughton encountered when he crossed the bar in 1792. It's now called Baker Bay. The mouth of the Columbia has been altered drastically since the days of Lewis and Clark. The Army Corps of Engineers has built jetties on both sides of the river that extend several miles out to sea. You, therefore, can no longer see the free-running Pacific from this spot. Don't fret, particularly if you have never before laid eyes on the Pacific. When you do finally see the ocean, you want it to be from the most spectacular site possible, and that is exactly what lies ahead.

Cape Disappointment

The Lewis and Clark Highway circles around Baker Bay to the quaint little fishing town of **Ilwaco.** The ridge you see off to the west is Cape Disappointment, named by the British fur trader Captain Meares who, on July 6, 1788, having learned from the natives about the existence of a river, tried but failed to cross the bar to escape a raging storm. Lewis and Clark spent their time here exploring the countryside and looking for a place for their winter camp.

November 17th, 1805 [Clark] The Chief of the nation below us came up to see us the name of the nation is Chin-nook *and is noumerous. they are well armed with good Fusees. I directed all the men who wished to see more of the Ocean to Get ready to set out with me on tomorrow day light.*

November 18th, 1805 [Clark] [Went] 2 Miles to the iner extremity of Cape Disapointment *passing a nitch in which there is a Small rock island, a Small Stream falls into this nitch from a pond which is ime-*

diately on the Sea coast passing through a low isthmus. this Cape is an ellivated circlier point covered with thick timber on the iner Side and open grassey exposur next to the Sea and rises with a Steep assent to the hight of about 150 feet above the leavel of the water

The U.S. Army converted this strategic spot into a coastal fortification, which it named Fort Canby, after the captain who was killed in the Modoc War. It is now a state park, with numerous overnight campsites. Turn left at the first intersection beyond Ilwaco's only traffic light, and follow the road along the water for three miles to the **Fort Canby State Park.** Then, instead of proceeding directly to the park headquarters, continue south along the narrow road that climbs to the **Lewis and Clark Interpretive Center** parking lot. A short trail leads up to the building, a portion of which was once a gun battery. Now go directly inside, proceed down the ramp, and take time to enjoy the exhibits describing the odyssey we have been following. The curators chose to focus on the scientific aspects of the trip, so you learn a bit about the practice of medicine, how the captains plotted their course, and what the Corps of Discovery had to eat. The museum is a bit like the one under the Arch in St. Louis, in that there are a number of large photographs showing the country Lewis and Clark saw on their westward journey. The St. Louis museum provided the syllabus for your trek west; this museum serves as a nostalgic reminder of all the incredibly beautiful scenery you saw along the way.

The planners designed the Lewis and Clark Interpretive Center so that your tour is climaxed by a walk up a ramp and into an enclosed observation room. *This is where you should get your first view of the Pacific Ocean.* The scene is breathtaking. The nearby lighthouse, built in 1850, is the oldest in the Pacific Northwest. Perched on a pillar of stone, it sends its feeble light 50 miles out to sea. The pounding waves, directly below, cast up giant spouts that are dashed against the cliffs, only to disappear in a cloud of frothy mist. Swells 20 feet high come roaring in at 30 miles an hour in the shipping lane between the jetties, breaking, reforming, and breaking again before the still waters of Baker Bay finally tame their fury. A freighter may be struggling to cross the bar. Woe to the captain who, in a gale, tries to fight against the tide. This is the famous Columbia Bar, graveyard of more than 200 major ships. Captain Meares's troubles were only the first. Jonathan Thorn, captain of the *Tonquin,* a 290-ton vessel with ten guns, tried to enter in 1811 and lost eight men in what was to be the first of the many disasters befalling his passengers, the Astorians. Thirty years later the *Peacock*, under the command of Captain Wilkes, went aground with such terrible loss that the United States Congress became convinced the Columbia River would never be much of a harbor and that Puget Sound was, therefore, to be the favored port in the Pacific Northwest. Indirectly, this led to the compromise that established the 49th parallel rather than 50°40' as the northern U.S. boundary. The bar is less dangerous now, thanks to the Army Corps of Engineers, but the Coast Guard has a lifeboat training school in the lee of the cape, just in case.

Poets have written that the loveliest place where land meets water is the Pebble Beach area of California's Del Monte Peninsula. Perhaps they should have visited Cape Disappointment first. Hollyhocks bloom in the sunnier spots while fir, pine, and sitka spruce cling tenaciously to the rocky cliffs. Driftwood in great profusion decorates the sand with nature's own pieces of sculpture. Bays and bayous are nesting grounds for a thousand birds. Underbrush, so thick and lush you feel you're in a rain forest, gives the whole area an enchanting, otherworldly flavor.

From Cape Disappointment, Captain Clark and his party of hearty souls walked north, looking for a place to build their winter headquarters. To trace their route, leave your car near the Park Headquarters and hike along the North Head Trail. Walking on this well-constructed trail, free of underbrush and fallen trees, is easy compared to what the explorers had to cope with.

November 13th, 1805 [Clark] I walked up the Brook & assended the first Spur of the mountain with much fatigue, the distance about 3 miles, through an intolerable thickets of Small pine. a groth about 12 or 15 feet high interlockd. into each other and scattered over the high fern & fallen timber, added to this the hills were so steep that I was compelled to draw my Self up by the assistance of those bushes.

The trail climbs to a windswept bluff, where the **North Head Lighthouse** occupies a commanding site. Captain Clark must have been standing here, for he later wrote:

November 19th, 1805 [Clark] after takeing a Sumptious brackfast of Venison which was rosted on Stiks exposed to the fire, I proceeded on through ruged Country of high hills and Steep hollers 5 miles. on a Direct line to the commencement of a Sandy coast which extended N. 10°W. from the top of the hill above the Sand Shore to a Point of high land

*distant near 20 miles. this point I have taken the Liberty of Calling after
my particular friend Lewis. I proceeded on the sandy coast 4 miles, and
marked my name on a Small pine, the Day of the month & year, &c.*

Long Beach

The sand shore Captain Clark was referring to is now the resort of
Long Beach, which the locals claim is the longest beach in the world.
Lewis Point, now called **Leadbetter Point,** is a bird sanctuary and nesting area for the Snowy Plover. All trace of the tree he carved his name
on has long since disappeared, so it's impossible to determine exactly
how far north he came. That's a pity, because Long Beachers claim that
this is the end of the Lewis and Clark Trail. They get some argument,
though, from others, who maintain that the Corps of Discovery's farthest advance down the Oregon coast should deserve that honor.

Long Beach's permanent population of 1100 soars considerably during the summer when vacationers come to swim, surf-fish, go clam digging, play golf, comb the beaches, or take the kids to the amusement
park. Both Oregon and Washington allow motorists to take their cars on
the beaches, but fine those who drive on clam beds or into the surf.
Willapa Bay, which separates Long Beach from the mainland, is the
largest oystering center on the west coast. Cranberries are grown farther
inland.

The Chinooks convinced the captains that the hunting would be better on the south side of the river:

November 24th, 1805 [Clark] They generaly agree that the Most Elk
*is on the Opposit Shore. added to this a convenient Situation to the Sea
coast where We Could make Salt, and a probability of Vessels comeing
into the Mouth of Columbia from whome we might precure a fresh Supply of Indian trinkets to purchase provisions on our return home; together with the Solicitations of every individual, except one of our party,
induced us to Conclude to Cross the river and if a Sufficent quantity of
Elk could probebly be precured to fix on a Situation as convient as we
could find.*

The fact that they put the question to a vote of the entire party was
highly unusual, considering that this was a military organization. Lewis
and Clark buffs have always been curious, but they will never know
who the lone dissenter was. The explorers met no ships while they were
at the mouth of the Columbia, although there is evidence that one sailed

into Baker Bay shortly after the party moved to Fort Clatsop. The historian Bernard DeVoto commented: "Jefferson's failure to send a ship to the mouth of the Columbia is inexplicable."

It wasn't that easy for the Corps of Discovery to get to the southern shore. Today it's a 5-minute drive and $2 to the toll collector, but for the Corps of Discovery it was 12 days before they settled in at what would become their home for almost four months. The weather was so bad and their canoes so unseaworthy, they had to retreat upriver 17 miles before they found water calm enough to cross. Most of that time was spent at **Tongue Point,** just east of present-day Astoria, Oregon.

Astoria

Astoria, named for John Jacob Astor, is the oldest town in the Pacific Northwest. The fort was established only five years after Lewis and Clark left for St. Louis. The settlement didn't have an auspicious beginning. First, two long boats and eight men were lost trying to brave the notorious Columbia Bar. The ship the Astorians arrived on, the *Tonquin,* sailed north shortly afterward to engage in fur trade on Vancouver Island. The careless captain, Jonathan Thorn, let too many tribesmen aboard. They promptly turned on the crew and massacred the lot save one, who, while hiding in the hold, set fire to the powder magazine, thereby blowing himself, the ship, and the captors to smithereens. Meanwhile, the overland party sent by Astor had such a time of it that few survived, and those who did lost heart for the business they were about. As noted earlier, when the British frigate *Beaver* trained its guns on the little fort during the War of 1812, the Astorians promptly sold the place, lock, stock, barrel, and fur, to the Northwest Company, which changed the name to Fort George.

The British didn't have a much better time of it. The tribes near The Dalles became increasingly hostile, making it difficult to get the furs to Fort George. In April of 1814 the frigate *Isaac Todd* arrived, bringing a new factor, Donald McTavish. Not wishing to brave the wilds of the Pacific Northwest without some civilization, he brought along his mistress, a voluptuous Portsmouth barmaid named Jane Barnes. She was the first white woman to come to this country. The chief of the local tribe could not understand why Jane refused his offer of what amounted to a "king's ransom" to induce her to his quarters. Perhaps the fact that he already had four wives had something to do with her decision. One of the more able employees of the firm, Alexander Henry, offered Jane his "protection," which no doubt made McTavish a bit angry. The two got to drinking and dared each other to row out to the *Isaac Todd,* which at that time was wallowing in waves as bad as those Jonathan Thorn faced in the *Tonquin.* The Northwest Fur Company lost two good men in that episode. Jane had seen enough of this wild country—she sailed on the *Isaac Todd* to the Orient, where she married an official of the British East India Company.

Shortly afterward, the Northwest Company was merged into the Hudson's Bay Company, and Fort George was abandoned in favor of Fort Vancouver. Later, Astoria prospered with an influx of salmon canneries that lasted until the catch dwindled in the 1960s. The port handled grain shipments as well as a huge export of forest products. Things are on the decline again, although the town does a fair tourist business. Maybe the geophysicists now scouting the area will find oil and things will boom again.

Astoria is a nice place to stay. While the motels are rather ordinary, there is a B&B located in an old convent uphill from the main street, and there are a number of good restaurants featuring the local catch. Showpiece of the town is the **Columbia River Maritime Museum,** probably the finest of its kind in the country. Among the displays are models of sailing ships, riverboats, and a full-size replica of the dugout our heroes used. Photos and artifacts from the numerous vessels lost on the bar illustrate why this has been called the graveyard of the Pacific. Docked outside is the lightship *Columbia,* which guided mariners across the bar for many years.

Astoria has an old-town atmosphere. Many of the houses are of the Victorian style usually associated with San Francisco. One of them, the **Captain George Flavel House,** is now an historical museum. **Astoria Tower,** a monument to Captain Gray, Lewis and Clark, the Astorians, and other early-timers, stands in a park at the top of a hill. A spiral staircase leads to a platform where you get a fantastic view of the bar, the harbor, and the lovely green mountains to the east.

Fort Clatsop

Lewis and Clark chose a site about 10 miles west of Astoria for their winter headquarters. Along the way, you pass by **Fort Stevens State Park,** site of an old military base that has the dubious distinction of being the only military establishment in the continental United States to have received hostile fire since the War of 1812. A Japanese submarine lobbed 17 5-inch shells at the fort, scoring a direct hit on the chain-link fence behind the batter's box at the baseball diamond. The campground here has more than 600 campsites.

Captain Lewis and an advance party arrived at the site of their winter quarters on December 7, 1805, and began construction of the fort, which, three days later, they named for the local Indian tribe. By Christmas eve they had moved in. Virtually the entire fort had disappeared by the time the Astorians arrived five years later. The present reconstruction was done in the 1960s by the Oregon Historical Society. The museum at Fort Clatsop is now operated by the National Park Service. There is a fine collection of artifacts associated with the expedition, including a replica of Lewis's air gun, muskets, peace pipes, tomahawks, leather clothes, and an example of the type of canoe the explorers used. A piece of sculpture depicts the two captains together with Lewis's dog, Seaman. Park rangers, dressed in buckskins, give lectures and demonstrations of the old-time techniques of muzzle-loading rifles, candle-making, and beaver-trapping.

Captain Clark's "O' the joy" mood soon changed to one of anger and despair, and his journal takes on a certain monotony. For days on end it begins with the same phrase: "Rained last night as usial." He was not the first nor the last explorer to question Balboa's choice of a name for this ocean.

December 1st, 1805 [Clark] The sea which is imedeately in front roars like a repeeted roling thunder and have rored in that way ever since our arrival in its borders which is now 24 days since we arrived in sight of the Great Western Ocian, I cant say Pasific as since I have seen it, it has been the reverse

Christmas day was no better.

December 25th, 1805 [Clark] at day light this morning we were awoke by the discharge of the fire arms of all our party & a Selute, Shouts and a Song which the whole party joined in under our windows. after brackfast we divided our Tobaccco which amounted to 12 carrots one half of which we gave to the men of the party who used tobacco, and to those who doe not use it we make a present of a handkerchief. The day proved Showerey wet and disagreeable.

During the 106 days the Corps of Discovery stayed here, it rained every day save 12. Clothing rotted and fleas infested the furs and hides of the bedding. The dampness gave nearly everyone rheumatism or colds. Yet, when it came time to leave, Captain Lewis reflected on the stay with a bit of stoicism:

> *March 20th, 1806 [Lewis] Altho' we have not fared sumptuously this winter and spring at Fort Clatsop, we have lived quite as comfortably as we had any reason to expect we should; and have accomplished every object which induced our remaining at this place except that of meeting with the traders who visit the entrance of this river.*

Three days later the explorers started on their long voyage home. Fort Clatsop, however, is not the end of the Lewis and Clark Trail. The Corps of Discovery desperately needed salt to cure and preserve their meat, so the Captains had some of the men establish a camp at present-day Seaside, Oregon.

Seaside

Seaside, Oregon's largest beach resort, is about 12 miles south of Fort Clatsop on U.S. Highway 101. The town used to have a large boardwalk, complete with rollercoaster, but it has been torn down and the place now has a somewhat more sedate atmosphere. Broadway, the main downtown street, terminates at the beach turnaround, which the Oregon Historical Society has designated as the official end of the Lewis and Clark Trail. It really isn't the end either, as we shall see. The 30-block-long residential area along the beach is dotted with fine old Cape Cod–style shingle-walled cottages. Signs point the way to the salt cairn, which is nestled between two lovely houses on Lewis and Clark Street. It's unlikely that Captain Lewis visited the site, but he did report on the progress.

> *January 5th, 1806 [Lewis] they commenced the making of salt and found that they could obtain from 3 quarts to a gallon a day; they brought with them a specemine of the salt of about a gallon, we found it excellent, fine, strong, & white; this was a great treat to myself and most of the party; I say most of the party, for my friend Capt. Clark declares it to be a mear matter of indifference with him whether he uses it or not.*

While at Fort Clatsop, the captains learned of the beaching of a whale near the salt works, so a group under the direction of Clark decided to go and have a look.

January 6th, 1806 [Lewis] Charbono and his Indian woman were also of the party; the Indian woman was very importunate to be permited to go, and was therefore indulged; she observed that she had traveled a long way with us to see the great waters, and that now that monstrous fish was also to be seen, she thought it very hard she could not be permitted to see either.

January 7th, 1806 [Clark] I proceeded on to near the base of a high Mountain where I found our Salt makers, the Salt Makers had made a Neet close camp, convienent to wood Salt water and the fresh water of the Clatsop river.

Ecola State Park/Cannon Beach

The high mountain Captain Clark was referring to was **Tillamook Head.** Lewis and Clark buffs will be delighted to know that a portion of the last segment of the Lewis and Clark Trail is, in fact, a trail. From the salt cairn, drive south on Beach Drive to Sunset Boulevard and continue on to the end of the road. Unlike Capt. Clark, you won't need a pilot; signs point the way.

January 7th, 1806 [Clark] I hired a young Indian to pilot me to the whale. we proceeded on under a high hill which projected into the ocian about 4 miles. my guide made a Sudin halt, pointed to the top of the mountain and made signs that we could not proceed any further on the rocks, but must pass over that mountain, I hesitated a moment & view this emence mountain the top of which was obscured in the clouds. I soon found that the path become much worst as I assended, and at one place we were obliged to Support and draw our selves up by the bushes & roots. after about 2 hours labour and fatigue we reached the top from which I looked down with estonishment to behold the hight which we had assended which appeared to be almost perpindicular.

The 6-mile-long trail to Ecola State Park climbs through about as pretty a forest as can be imagined. The trees here are nothing like any Captain Clark had seen back east. He was astonished to note that "Some Species of pine or fir rise to the emmence hight of 210 feet and from 8 to 12 feet in diameter, and are perfectly sound and Solid." The country is vastly different from the plains. Here in this dense forest of Douglas fir and sitka spruce, instead of "Big Sky Country" you have no sky at all. Moss- and lichen-covered limbs capture the constantly swirling fog, dripping the condensate onto Oregon grapevines, sword-ferns, and pink and white rhododendrons. Boggy portions of the trail are paved with small tree trunks, laid cross-wise like an old-time cordwood road. The mist and the gloom combine with a delicacy of colors to make this truly an enchanted forest.

That night the party camped near the summit of Tillamook Head, where you have a lovely view of the ocean and the now abandoned lighthouse. The next morning was apparently clear.

> *January 8th, 1806 [Clark] from this point I beheld the grandest and most pleasing prospect which my eyes ever surveyed, in my frount a boundless Ocean; to the N. and N.E. the coast as far as my sight could be extended, the Seas rageing with emence waves and breaking with great force*

That day, the Corps of Discovery "proceeded on" as far as they would go.

> *January 8th, 1806 [Clark] Wind hard from the S.E. and See looked wild breaking with great force against the Scattering rocks and the ruged rockey points under which we wer obleged to pass and if we had unfortunately made one false Step we Should eneviateably have fallen into the Sea and dashed against the rocks in an instant, fortunately we passed over 3 of those dismal points and arived on a butiful Sand Shore and proceeded to the place the whale had perished, found only the Skelleton of this Monster the Whale was already pillaged of every Valuble part. this Skeleton measured 105 feet. I returned to the Village on the creek which I shall call E co-la or Whale Creek.*

Ecola Creek separates Tillamook Head from the town of Cannon Beach. Those not wishing to make the 6-mile trek over the top can drive into Ecola State Park from the south. From the parking lot a short pathway leads to a grassy bluff above the crashing waves. The view is south, toward Haystack Rock. To paraphrase Captain Clark, you will behold "the grandest most pleasing prospect which your eyes will ever survey." It's the most photographed scene in Oregon.

Cannon Beach is a lovely seaside resort that in recent years has taken on a "whaling village" atmosphere. The main street is now refurbished with rustic buildings housing art galleries, boutiques, and dozens of restaurants and coffee houses.

So, we have finally reached the end of the Lewis and Clark Trail. Or have we? Keep in mind Jefferson's instructions:

Should you find it safe to return by the way you go, do so; On reentering the United States and reaching a place of safety, discharge your attendants and repair yourself, with your papers, to the seat of government.

That's exactly what they did. The end of the Lewis and Clark Trail is in St. Louis.

PRACTICAL ADVICE

It's appropriate to give some advice on the nitty-gritty aspects of following the Lewis and Clark Trail. The Corps of Discovery stayed wherever it chose, stopped when tired, and lived off the land—but you can't. You're constrained by having to go where the roads go and stay where the motels or campgrounds are, as well as by the need to respect private land. Even those intrepid captains felt it necessary to hole up for the winter, and you should, too. Listed below are some aspects of traveling along this particular trail that you should keep in mind.

WHEN TO GO As we have seen, there are thousands of things to see and do along the Lewis and Clark Trail, but some are only possible during the summer months. The numerous little museums that tell so much about the history and character of an area are open, for the most part, only between Memorial Day and the end of September. Parts of Yellowstone and Glacier Parks are open year-round, but the great old-time resort hotels, which rely on college students for their help, only open their doors between mid-June and early September. Those who put a priority on visiting the mountain passes the explorers crossed may have to wait until the first of August because it usually takes that long for the snow to melt from the high country. White-water enthusiasts, by contrast, will want to get an early start, because that's when the rivers run high and the sport is more exhilarating. Fishermen and hunters, of course, have their own favorite times. We visit some off-the-beaten-track places in the Plains States, where the roads are not likely to be plowed, so even there it's best to wait at least until April.

The question always comes up of whether you should have a fixed itinerary or just go when and where the spirit moves. That's a matter of personal preference, but those with a fixed schedule should realize that the weather may be terrible on the day you're planning to walk the Eagle Creek trail, for example, so you should have some contingency plans.

MOTELS Except for places of special interest, I have not felt it necessary to describe accommodations along the Lewis and Clark Trail. Those sections of the route that are near an interstate or a major U.S. highway have a plethora of motels, many of which belong to chains. Most offer toll-free reservation numbers, so you can book ahead if you're going to arrive late. Virtually all the little off-the-beaten-track towns we visit have one or more motels, some of which have a uniqueness not found in those along the more well-traveled routes. In the course of the research for this book (which, incidentally, was done at the height of the summer season), I stayed at about 50 different motels. I never once felt the need to make a reservation and was never turned away, although I did take the precaution of quitting the road by six in the evening. Only in Missoula, Montana, when several thousand motorcycle enthusiasts were having a convention, did I see the NO VACANCY lights go on, and then only after about eight in the evening. There are so many motels these days, my advice is not to worry.

BED AND BREAKFAST More and more travelers are beginning to discover the delights of staying in private homes. It's an escape from the sameness of motels. You not only get a chance to peek in at how other people live, but also learn much more about the country than you otherwise would. The hostess (most, but not all, B&Bs are run by women) is usually an expert on unusual things to see and do, and often knows a lot about the local history and lore. Most places that open their doors to travelers are charming old homes filled with period furniture. Breakfast is a chance to sample some real down-home cooking.

You'll find most of the B&Bs near the cities or in resort areas. Western Montana, for example has about 25. They aren't always easy to find. Most don't advertise widely for fear of attracting an undesirable clientele. For this reason, many have banded together to form an association, often appointing one member to act as a central booking agent. An example is the Missouri Bed & Breakfast Association (P.O. Box 31246, St. Louis, MO 63131). About a dozen books describing B&Bs are available at your local bookseller, but I haven't found any that list more than a couple of places along the Lewis and Clark Trail. While last-minute accommodations can be found, most B&B operators prefer to have people book ahead. In spite of these obstacles, my experience is that the rewards are well worth the extra effort.

CAMPGROUNDS Lewis and Clark followed the Missouri, Salmon, Bitterroot, Clearwater, Snake, and Columbia rivers, all of which are now playgrounds for fishermen and other watersport enthusiasts. As a result, the trail is liberally dotted with campgrounds. The State of Missouri alone has about 25 parks near the river that permit overnight camping. Kansas, Iowa, and Nebraska, between them, have about 25 more. Once past Sioux City, Iowa, you're in the domain of the U.S. Army Corps of Engineers, which has built hundreds of campgrounds along the newly formed lakes. Most of them have hook-ups for RVs. The portions of the trail that traverse the Rocky Mountains and the Cascades go through national forests, which, of course, have campgrounds, too. The three National Parks we visit—Theodore Roosevelt, Yellowstone, and Glacier—all have campsites administered by the National Park Service. The upper Columbia River is Corps of Engineers country, so again you find lakeside parks, and the states of Oregon and Washington have provided numerous campsites along the lower river. All told, there are probably 150 campgrounds along the route; an average of 1 every 25 miles.

FOOD The Lewis and Clark Trail goes primarily through "steak and potatoes" country, so, except in the big cities, don't expect to find out-of-the-ordinary places to eat. Seafood restaurants exist for the most part only between Portland and the ocean. Many towns have a "Chinese" restaurant, but here again, they're the chop suey type; you won't find dim sum and won't be able to savor the delights of Cantonese cooking. Counterbalancing this, you'll be delighted to find that most restaurants feature home-style cooking where, often, the chef is also the owner and the staff has worked there for years. Much of our route is away from the traditional tourist paths, so you won't find many chain restaurants, except, of course, for the fast-food outlets, which are as prevalent here as anywhere else in the country. Almost every little town has a downtown coffee shop; if typical, it has 10 booths and a 20-stool counter. This is where the local people come to meet their friends and pass along the local gossip. BLT sandwiches and soup-of-the-day are the favorite luncheon fare. What impressed me most about these places is that some of them still serve the milk for your coffee in those tiny little glass bottles, which, inevitably, the waitress perches on the side of the saucer. For us city dwellers, that little symbol, a throwback to earlier times, says more about what this country is like than almost anything I can think of.

DRINK Bottled liquor is sold in state-operated stores in every state we pass through. They all close on Sundays and holidays. Liquor stores in Montana also close on Mondays, but bars, which are open seven days a week, sell off-sale bottles at a slightly higher price. Liquor by the drink is available in all states except Kansas, where you have to join a "private club" before you can imbibe in public. Some localities in Nebraska are dry. On Sundays bars don't open until noon in Iowa and Kansas, 1 P.M. in Missouri, South Dakota, and Montana, and at 6 P.M. in Nebraska.

HIGHWAYS The interstate highway system has made travel in the western United States infinitely more pleasurable. All those people who are in such a hurry and all those big 18-wheel trucks that used to clog the little roads are now on the interstates. The U.S. highways, formerly the backbone of the motoring infrastructure, are now almost deserted. Driving the backroads now seems like a dream come true. You look forward to the next town, for it's a chance to slow down, look around a bit, and perhaps get out and stretch your legs with a stroll down Main Street. Only a few sections along the Lewis and Clark Trail require driving on an interstate, and for much of the trip the nearest one is several hundred miles away. The result is that you get a much better feeling for what this great country is like because things take on a more local character.

Some of the trek follows state highways and county roads, which, for the most part, are well maintained. The main hazards are farmers driving their tractors along the shoulder and ranch hands herding their cattle from one range to another. Even here, though, you can average 50 or more miles an hour if you choose. We do follow some gravel and dirt roads, which, except as noted, are generally good; if you don't want to get your car all dusty, you can avoid them without missing too much. At the risk of being simplistic, I can definitely state that the smaller the road you travel, the more fun you'll have.

AUTO REPAIR The country between Omaha and Portland is the land of the American automobile. Less than 10% of the cars you see are foreign-built. This suggests that if you are driving a non-U.S. car, you should have it serviced before you leave. I went to about 10 garages and had to backtrack 150 miles before I found someone who would work on my 10-year-old Volvo.

SOURCES OF INFORMATION We pass through 11 states on our way to the Pacific: Illinois, Missouri, Kansas, Iowa, Nebraska, South Dakota, North Dakota, Montana, Idaho, Washington, and Oregon. Each has an office to promote tourism in the state. Write to the Office of Tourism at the state capitol. You'll receive, free, a state road map and a booklet describing things to see and do, and a list of hotels and motels. Most booklets also list the dates of festivals, parades, rodeos, and county and state fairs. Many of the states also operate tourist information centers along the interstate highways. The people who staff these centers are always well informed and have up-to-date literature.

Most towns and cities we pass through have a Chamber of Commerce that distributes free information on things of local interest. Also, the hundreds of museums we visit along the way are gold mines for local literature and lore, and many have well-stocked bookstores.

Information available at your local bookstore includes: *Mobile Travel Guide, AAA Travel Guide, Woodall's Campground Directory,* and *Rand McNally Campground and Trailer Park Directory.*

Readers who want to learn more about our heroes should by all means join the Lewis and Clark Trail Heritage Foundation, which publishes the quarterly journal *We Proceeded On.* The address is 5054 S.W. 26th Place, Portland, Oregon 97201.

PARTING WORD It is 2000 miles between St. Louis and Portland via the interstates that roughly parallel the Oregon Trail; a 4-day drive for those who don't want to spend more than ten hours a day behind the wheel of a car. Lewis and Clark, as we have seen, took a more circuitous route, roughly 750 miles longer. You can still drive it in 6 days. The captains took alternate routes home, though, which adds another 1000 miles to the trek, but if you follow these routes, too, you'll do a lot of backtracking. It is still, however, possible to see the entire trail in 10 days. Don't try. Two months is about the right amount of time to take to really see this country. Most of us don't have the luxury of being away from home and job for that long, but, nevertheless, the best advice I can give is simple—don't hurry.

BIBLIOGRAPHY

Adventures on the Columbia River
Ross Cox
Binfords & Mort
Portland, OR

Bury My Heart at Wounded Knee
Dee Brown
Holt, Rinehart & Winston
New York, NY (1970)

Coyote Was Going There: Indian Literature of the Oregon Country
Compiled and edited by Jarold Ramsey
University of Washington Press
Seattle, WA (1977)

Exploration & Empire
William H. Goetzmann
W.W. Norton & Company
New York, NY (1966)

History of the Expedition Under the Command of Lewis and Clark
Elliott Coues and Francis P. Harper
New York, NY (1893)
Reprinted by Dover Publications, Inc.
New York, NY

The Journals of Lewis and Clark
Edited by Bernard DeVoto
Houghton Mifflin Company
Boston, MA (1953)

The Journals of The Expedition Under the Command of Capts. Lewis and Clark
Edited by Nicolas Biddle
Bradford and Inskeep; and Abm. H. Inskeep
New York, NY (1814)
Numerous reprintings

Lewis and Clark: Historic Places Associated with Their Transcontinental Exploration (1804–06)
Roy E. Appleman
United States Department of the Interior
National Park Service
Washington, DC (1975)

Lewis and Clark Partners in Discovery
John Bakeless
William Morrow and Co., Inc.
New York, NY (1947)

Lewis and Clark: Pioneering Naturalists
Paul Russell Cutright
University of Illinois Press
Urbana, IL (1969)

Manuel Lisa and the Opening of the Missouri Fur Trade
Richard Edward Oglesby
University of Oklahoma Press
Norman, OK (1963)

The Men of the Lewis and Clark Expedition
Charles G. Clarke
The Arthur H. Clark Co.
Glendale, CA (1970)

Meriwether Lewis: A Biography
Richard Dillon
Coward-McCann, Inc.
New York, NY (1965)

Only One Man Died: The Medical Aspects of the Lewis and Clark Expedition
E.G. Chuinard, MD
The Arthur H. Clark Co.
Glendale, CA (1979)

Original Journals of the Lewis and Clark Expedition
Reuben Gold Thwaites
Dodd, Mead & Company
New York, NY (1905)

Stern Wheelers up Columbia
Randall V. Mills
University of Nebraska Press
Lincoln, NB (1947)

The Oregon Trail Revisited
Gregory M. Franzwa
Patrice Press, Inc.
St. Louis, MO (1972)

We Proceeded On
The official Publication of the Lewis and Clark Trail Heritage Foundation, Inc.
Robert E. Lange, Editor
5054 S.W. 26th Place
Portland, OR 97201

Views of a Vanishing Frontier
John C. Ewers et al
University of Nebraska Press
Lincoln, NB (1984)

INDEX

Absaroka Wilderness Area, 150
Absarokee, MT, 150
Agricultural Hall of Fame, 37
Albrecht Art Museum, 42
Alder Creek, 161
Alder Gulch, 162
Alder, MT, 164
Alexander, ND, 96
Alice Creek, 201
Alpowai Interpretive Center, 230
Altoona, WA, 272
American Fur Company, 86, 101, 113, 115
Anaconda Mine, 159
Anaconda, MT, 130, 161
Arrow Rock, MO, 17, 28, 29
Arrowsmith, Aaron, 82
Ashley, General William, 72, 101, 141
Astor, John Jacob, 86, 235, 268, 279
Astoria Tower, 280
Astoria, OR, 279
Astorians, 222, 233
Atcheson, KS, 39
Atcheson, Senator David, 39
Atkins Museum, MO, 36
Atkinson, General Henry W., 53
Audubon, John J., 70, 92, 101
Augusta, MO, 19

Bad River, SD, 66, 67
Bainville, MT, 104
Bannack City, MT, 162, 181
Bannack State Park, MT, 181
Bannock Pass, 169
Baptiste (Pomp), 10, 86, 89, 144
Barlow Pass, 250
Barlow, Sam, 242

Barnes, Jane, 279
Battle Ground, WA, 270
Battle Mountain, 183
Beacon Rock, 260
Bear Tooth Mountain, 146, 147
Bear Tooth Plateau, 147
Bear's Paw Mountains, 111, 121
Beaverhead County Museum, 165
Beaverhead National Forest, 164
Beaverhead River, 164
Beaverhead Rock, 164
Beaverhead Valley, 158
Becknell, Wm., 29
Beckworth, Jim, 72, 135, 141
Bellefontaine Cemetery, 15
Bellevue, NB, 48
Bennett Pass, 250
Benson, Simon, 249, 266
Benteen, Capt., 142
Benton, Senator Thomas Hart, 15, 115
Berkeley Pit Mine, 160
Biddle, Henry J., 260
Biddle, Nicolas, 1
Big Belt Mountains, 122
Big Bend Dam, 66
Big Hole National Battlefield, 183
Big Hole River, 183
Big Horn River, 139–143
Big Lake Park, IA, 51
Big Sky, MT, 154
Big Timber, MT, 150
Biggs, OR, 240
Billings, Fredrick, 153
Billings, MT, 145, 146
Bingham, George Caleb, 33
Bingham-Waggoner Estate, 33
Bismarck, ND, 80, 81

Bismarck, Otto von, 80
Bitter Creek Valley, 144
Bitterroot Mountains, 178, 199
Bitterroot River, 176
Bitterroot Valley, 177
Black Eagle Falls, 126
Blackfeet Indian Reservation, 191
Blackfeet River, 201
Blair, NB, 54
Bodmer, Karl, 49, 50, 70, 88, 96, 115, 118
Bonner Springs, KS, 37
Bonneville Dam, 249, 255–257
Bonneville, Capt. B.L.E., 235, 255, 269
Boone's Lick, 28
Boone's Lick Trace, 17
Boone, Daniel, 18
Boone, Daniel Morgan, 17, 32
Boonville, MO, 28
Boot Hill, 163
Bozeman Pass, 152
Bozeman, John, 153
Bratton, Wm., 89, 143
Bridge of the Gods, 254
Bridger, Jim, 33, 72, 101, 135, 153
Broughton Flume, 251
Broughton, Lt. Wm. R., 235, 245, 275
Brown, Dr. Barnum, 109
Browning, MT, 192, 193
Brownville, NB, 45
Bullwhacker, Coulee, 121
Burke Act of 1906, 55
Butte City, MT, 130, 159, 161
Butterfield, D. A., 39
Bybee-Howell Museum, 270
Bynum, MT, 197

Calamity Jane, 152
Camas Prairie, 216, 221, 227
Cameahwait, 166, 172
Camp Disappointment, 191, 192
Camp Fortunate, 166
Campbell, Robert, 141
Cannon Beach, OR, 284
Cape Disappointment, 275–277
Carroll College, 131
Carson Hot Spring, 250
Carson, Kitt, 135

Carson, WA, 261
Cascade Falls, 244, 254
Cascade Locks Museum, 254
Cascade Locks, OR, 252–254
Cathedral of St. Helena, 131
Cathlamet, WA, 271
Catlin, George, 70, 88, 96, 101, 126, 129
Celilo Falls, 241
Chamberlain, SD, 63–65
Channing, Reverend George, 57
Charboneau (Carbono) (Shabono), 10, 85, 89, 90, 108, 127, 143, 155, 176, 283
Chardon, Francis, 88
Cheyenne River Indian Reservation, 59
Chico Hot Springs, 149
Chief Blackbird, 56
Chief Crazy Horse, 100
Chief Crow King, 100
Chief Four Bears, 93, 96
Chief Gall, 76, 100
Chief Grey Eyes, 72
Chief Joseph, 100, 122, 147, 182, 184, 206, 227
Chief Joseph Battleground, 110
Chief Joseph Pass, MT, 186
Chief Kamiakin, 259
Chief Looking Glass, 110, 184
Chief Red Cloud, 38, 153
Chief Sitting Bull, 73, 75, 76, 100, 110, 143
Chief Smutty Bear, 62
Chief Timothy Park, 230
Chief Twisted Hair, 220
Chief White Bird, 227
Chinook, MT, 110
Choteau, MT, 198
Chouteau, Auguste, 11, 69, 110, 141
Chouteau, Francois, 35
Chouteau, Pierre, 11, 141
Chouteau, Pierre, Jr., 69
Citadel Rock, 118
Clackamas County Museum, 267
Clark Canyon Dam, 166
Clark County Museum, 269
Clark Fork River, 159
Clark's Fork Yellowstone, 141, 147
Clark, George Rogers, 56

Clark, William A., 130, 159
Clarkston, WA, 229
Clay County Museum, 33
Clearwater Battle, 216
Clearwater National Forest, 210, 222
Clearwater River, 184, 216, 223
Clearwater/Snake River Confluence, 228
Clyman, James, 141
Coal Banks Landing, 117
Collins, Pvt. John, 37, 86, 89
Colter Falls, 127
Colter Pass, 147
Colter's Hell, 10, 141
Colter, John, 20, 86, 89, 132, 135, 141
Columbia Gorge, 251–261, 263
Columbia Gorge Hotel, 249
Columbia Plateau, 238, 239
Columbia River, 220
Columbia River Maritime Museum, 280
Columbus, MT, 150
Comstock, Henry T. P., 153
Continental Divide, 169, 170, 175, 185, 199, 200
Cook, C. W., 148
Cooke City, WY, 147
Cooke, Jay, 80
Cooke, Jay, Jr., 147
Copper King Mansion, 160
Copper Village Museum, 161
Corn Palace, SD, 64
Coues, Eliott, 2
Council Bluffs, IA, 42, 51–53
Cow Creek, 122
Cox, Russ, 242
Cramer-Kenyon Heritage House, 62
Crooked Falls, 127
Cross Ranch Nature Preserve, 81
Crow Agency, 143
Crow Creek Indian Reservation, 65, 143
Crown Point, 258
Cruzette, Peter, 84, 90, 101
Curtis, Genl. Samuel R., 35
Custer Battlefield, 142, 143
Custer, Elizabeth, 78
Custer, Lt. Col. George Armstrong, 76, 142, 143

Cut Bank Creek, 190, 197
Cut Bank, MT, 112, 190
Cutright, Dr. Paul, 69

Dakota Territorial Museum, 62
Darby, MT, 177
Dauphine Rapids, 120
Dawes Act of 1887, 55
Day, John, 239
Dayton, WA, 231
de Montigny, Dumont, 28
De Smet, Father, 53, 76, 101, 102, 179
Dearborn River, 128, 198, 199
Deer Lodge, MT, 161
Dempsey, Jack, 190
Deschutes River, 240
DeSoto Bend National Wildlife Refuge, 53
DeVoto Memorial Grove, 211
DeVoto, Bernard, 279
Dillon, MT, 165, 181
Dred Scott, 14
Drouillard (Drewyer), George, 22, 29, 89, 127, 132, 135, 140, 169, 190, 194, 196, 205, 239
DuBois River, IL, 9

Eads Bridge, 13
Eads, William, 182
Eagle Creek, 255
Earhart, Amelia, 39
East Alton, IL, 9
East Glacier, MT, 193, 195, 199
Eastern Montana College, 145
Ecola State Park, 283, 284
Edgar, Henry, 161
Edgerton, Sidney, 182
Elkhorn Hot Springs, 182
Elkhorn Mountains, 158
Ennis, MT, 164
Essex, MT, 195
Evans, Pvt. John, 11, 89, 176
Ewing, General Thomas, 35
Excelsior Springs, MO, 33

Fairweather, Bill, 161
Femme Osage Creek, 18
Fetterman, Lt. Col. Wm., 153

Fields, Joseph, 40, 89, 99, 190, 196, 198, 205, 219
Fields, Reuben, 89, 190, 196, 205, 219
Fight Site, 191, 195–197
Fitzpatrick, Wm. (Broken Hand), 72
Flathead Mountains, 199
Flathead River, 195
Flathead Valley, 204
Flavel, Capt. Geo. Museum, 280
Floyd, Sgt. Charles, 9, 56
Folsom, David, 148
Ford, Bob and Charles, 42
Forest Park, MD, 14
Fort Abraham Lincoln State Park, ND, 76, 78
Fort Assiniboine, 111
Fort Astoria, 268, 269
Fort Atkinson, 45
Fort Belknap Indian Reservation, 110
Fort Benton Museum, 116
Fort Benton, MT, 109, 115, 116, 199
Fort Berthold Indian Reservation, 92
Fort Buford, 99
Fort Canby State Park, 276
Fort Cascades, 259
Fort Clark, 86
Fort Clatsop, 280–282
Fort Columbia State Park, 274
Fort Fizzle, 206
Fort George, 279, 280
Fort Keogh, 139
Fort Lapwai, 227
Fort Leavenworth, 111
Fort Mandan, SD, 82–86
Fort McKenzie, 50, 113
Fort Missoula, 203
Fort Osage, 30, 31
Fort Owen, 179
Fort Peck Dam, 108
Fort Peck Indian Reservation, 104
Fort Peck Tribal Museum, 104
Fort Piegan, 113
Fort Pierre, 49, 70
Fort Randall Dam, 63
Fort Raymond, 140
Fort Rice, ND, 76
Fort Simcoe, 238
Fort Spokane, 269
Fort Stevens State Park, 280
Fort Thompson, SD, 65
Fort Union, 49, 75, 101, 113, 137
Fort Vancouver, 256, 259, 269, 280
Fort Walla Walla, 269
Fort Walla Walla Museum, 233
Four Bears Bridge, 93
Frazier, Pvt., 205, 215
Fremont, John C., 261
Frey, Jonnie, 41
Frontier Gateway Museum, MT, 138

Gallatin River, 134, 135, 152, 153
Gardiner, MT, 148
Garrison Dam, 91
Gasconade River, 21
Gass, Sgt. Patrick, 10, 60, 83, 89, 205, 223
Gates of the Mountains Wilderness, 129
Gates of the Rock Walls, 117–122
Gates of the Rocky Mountains, 128, 129
Gavins Point Dam, 62
Ghost Dance, 73
Giant Springs, 126, 127
Gibbon, Col. John, 184
Gibbons Pass, 185
Gibbons, Tommy, 190
Gibson, Pvt. George, 89, 149
Glacier National Park, 112, 191–195
Glacier Park Lodge, 193
Glasgow, MT, 110
Glass, Hugh, 72
Glendive, MT, 137
Going to the Sun Highway, 195
Goldendale, WA, 238
Goodrich, Pvt., 205
Grand Coulee Dam, 237
Grand Tetons, 141
Grand Union Hotel, 116
Grangeville, ID, 217
Granite Peak, 147
Grant, U. S., 142, 269
Grant-Kohrs Ranch National Monument, 161
Grasshopper Creek, 116, 181
Gravelly Range, 163
Gray's Bay, WA, 245
Gray, Capt. Robert, 145, 268

Index 299

Great Falls of the Missouri, 122–125
Great Falls, MT, 125–127, 198
Great Plains Black Museum, NB, 50
Great Plains Museum, SD, 78
Green Mountain, 201
Greer, ID, 218
Greycliff Prairie Dog Town State Monument, 151

Haley, Captain, 275
Hall of Waters, MO, 33
Hall, Pvt. Hugh, 37
Hamilton Museum, 178
Hamilton, MT, 178
Hancock, General Winfield Scott, 99
Hancock, Julia (Mrs. Wm. Clark), 120
Hanford Science Museum, 236
Hardin, MT, 144
Hardscrabble Creek, 200
Harney, Col. Wm. S., 70
Havre, MT, 187
Hayden, Ferdinand, 148
Haystack Rock, 284
Hayworth Park, 48
He Devil Peak, 228
Hearst, George, 159
Hearst, Phoebe Apperson, 161
Hearst, Wm. Randolph, 161
Helena, MT, 129–131
Hell Roarin' Gulch, 160
Hellgate, 201, 202
Hells Canyon, 228, 229
Henry Doorly Zoo, NB, 51
Henry, Alexander, 279
Henry, Andrew, 31, 72, 101, 141
Hensler, SD, 82
Hermann, MO, 20
Hill, James J., 97, 110, 240
Hill, Sam, 240
Holladay, Ben, 37, 39, 41
Homestead Act of 1862, 98
Hood River, 249
Horsetail Falls, 258
Howard, General O. O., 110, 182, 184, 206, 216, 227
Howard, Pvt. Thomas, 89
Hudson's Bay Company, 82, 101, 172, 177, 235, 267, 268
Hunt, Wilson Price, 47, 235, 239

Ilwaco, WA, 275
Immel, Michael, 141
Independence Creek, 39
Independence, MO, 32, 33
Index Peak, 147
Indian Post Office, 213
Indian Springs, 174
Iowa, Sac, Fox Indian Reservation, 43
Irving, Washington, 16, 222, 256, 264

Jackson County Museum, 36
Jackson's Hole, 141
Jackson, Andrew, 255
Jackson, David E., 141
Jackson, MT, 183
James Kipp State Park, 109, 117, 122
James, Frank, 32, 55
James, Jesse, 33, 35, 42
Jefferson Barracks Historic Park, 15
Jefferson City, MO, 21
Jefferson Landing, MO, 22
Jefferson Memorial (Gateway Arch), 13
Jefferson River, 133, 164
Jefferson, Thomas, 2, 121
Jerry Johnson Hot Springs, 213
Jim Bridger Ski Resort, 154
John Brown Cave, NB, 47
John Day Dam, 239
Jones, Robert, 141
Joslyn Art Museum, 49, 50
Judith Basin, MT, 139
Judith River, 120

Kamiah, ID, 217
Kane, Paul, 261
Kansas City, MD/KS, 34–37
Kearney, General, 15
Kemper Military School, 28
Kennewick, WA, 236
Kildeer, ND, 92
King Limhi, 172
Kipp, James, 88, 101, 113
Klein Museum, 73
Knife River Indian Village, 88
Kooskia, ID, 216

La Page, Baptiste, 90
Labadie, Sylvester, 31, 141

Labiche, John, 9, 89, 96, 176
Laclede, Pierre Liquest, 11
Lake McDonald Lodge, 193
Lake Sakakawea, 92
Lancaster, Samuel C., 248
Langford, N. P., 148
Larch Mountain, 259
Last Chance Gulch, 130
Latourell Falls, 258
Laurel, MT, 146, 149
Leadbetter Point, WA, 278
Leavenworth, Col. Henry, 72
Leavenworth, KS, 38, 39
Lee, Daniel, 244
Lemhi Pass, 166–171
Lemhi River, 171
Lewis and Clark Caverns, 155
Lewis and Clark College, 266
Lewis and Clark Interpretive Center, WA, 276
Lewis and Clark Memorial Park, MT, 104
Lewis and Clark Monument, MT, 192
Lewis and Clark Pass, 200, 201
Lewis and Clark Salt Carin, 282
Lewis and Clark State Park, IA, 54
Lewis and Clark State Park, IL, 7
Lewis and Clark State Park, MO, 40
Lewis and Clark State Park, WA, 232
Lewis River, 262
Lewis, Reuben, 31, 141
Lewiston, ID, 180, 228
Lexington Battlefield, 30
Lexington, MO, 29, 30
Liberty Memorial, MO, 36
Liberty, MO, 33
Libiche, Francis, 90
Like a Fishhook Village, 92
Lincoln, MT, 201
Lisa, Manuel, 12, 15, 31, 40, 101, 140, 141
Little Belt Mountains, 122
Little Dixie, MO, 25–32
Little Missouri Bay State Park, ND, 92
Livingston, MT, 151
Lochsa River, 211–215
Loess Bluffs, MO, 45
Logan Pass, 195
Logan, MT, 154

Loisel, Regis, 66
Lolo (Travelers Rest), MT, 180, 186
Lolo Campground, 220
Lolo Hot Springs, 207
Lolo Motorway, 210
Lolo Pass, 207, 210
Loma, MT, 112
Lonesome Cove, 213
Long Beach, WA, 278
Long Camp, 218
Lost Trail Pass, 175, 185
Lowell, ID, 216
Lower Brule Indian Reservation, 59
Lyceum Theatre, 29
Lyon, Capt. Nathaniel, 28

MacDonald, Finnan, 172
Macy, NB, 55
Madison Buffalo Jump State Park, 154
Madison County Museum, MT, 163
Madison River, 133
Malmstrom Air Force Base, 124
Mammoth Hot Springs, 148
Mandan Villages, 82
Mandan, ND, 78
Many Glacier Lodge, 193
Marias Pass, 195
Marias River, 112, 187, 189
Marion, MO, 23, 24
Marmaduke, Col. John L., 28
Marsh, Capt. Grant, 141, 144
Marshall, James W., 41
Marshall, MO, 29
Marthasville, MO, 18
Maryhill, WA, 239
Matson, MO, 18
Maxey, Dr. Chester, 234
Maximilian, Prince, 49, 70, 84, 86, 88, 96, 101, 115, 118
McCord Creek Falls, 257
McGhee, Col. James H., 35
McKenzie, Alexander, 84, 121
McKenzie, Donald, 235
McKenzie, Kenneth, 101
McLaughlin House Museum, 267
McLaughlin, Dr. John, 268, 269
McLean County Historical Society, ND, 83
McNary Dam, 235

McNeal, Pvt. Hugh, 89, 169, 170, 205
McTavish, Donald, 279
Mears, Capt., 275, 276
Megler, WA, 273
Memalose Rest Area, 248
Menard, Pierre, 31, 135, 140
Mengarini, Gregory, 179
Meriwether, MT, 192
Miles City, MT, 139
Miles, Col. Nelson (Bear Coat), 110, 139
Milk River, 105, 187
Miller, Joaquin, 221
Milltown, MT, 201
Minot, ND, 74
Missoula Art Museum, 203
Missoula, MT, 180, 199, 202, 203
Missouri Botanical Garden, 15
Missouri Buffalo Jump, 120
Missouri Fur Company, 30
Missouri Headwaters State Park, 133
Missouri Historical Society, 14
Missouri Manifesto, 32
Missouri River Outfitters, 117
Missouri Town 1855, 32
Mitchell, SD, 64
Mobridge, SD, 71
MonDak Heritage Center, 137
Monida, MT, 170
Montana School of Mines, 160
Montana State Museum, 131
Montana State University, 153
Mormon Center, MO, 33
Morrison, William, 140
Morton, J. Sterling, 45
Moutain Goat Overlook, 195
Mt. Adams, 238, 249, 254
Mt. Helena, 131
Mt. Hood, 235, 238, 245, 254, 259
Mt. Jefferson, 264
Mt. Ranier, 238
Mt. St. Helens, 238, 245, 254, 261, 262
Mt. St. Helens National Volcanic Monument, 261, 262
Mullan Road, 199, 202, 232
Mullan, Lt. John, 116, 199, 210
Mulligan, Col. James A., 30
Multnomah County, 264

Multnomah Falls, 258
Museum of Montana Wildlife, 193
Museum of the Plains Indians, 193
Museum of the Rockies, 154

Naselle, WA, 273
National Bison Range, 204
National Museum of Transport, 15
Nelson Gallery of Art, MO, 36
Nevada City, MT, 163
New Franklin, MO, 28
New Market, 266
New Town, ND, 91
Newman, Pvt. John, 75
Nez Perce Indian Reservation, 216, 226–228
Niobrara, NB, 62
North Dakota Heritage Center, 81
North Dakota Zoo, 81
North Fork, ID, 174
North Head Lighthouse, 277
Northern Montana College, 111
Northwest Fur Company, 82, 235, 268, 279

Oahe, Lake, SD, 70
Offutt Air Force Base Museum, NB, 48
Old Governors Mansion, MT, 131
Old North Trail, 197
Omaha Indian Reservation, 55
Omaha, NB, 48–51
Oneonta Falls, 258
Ophir, MT, 115
Ordway, Sgt. John, 9, 89, 137, 199
Oregon City, OR, 265, 267
Oregon Historical Society Museum, 266
Oregon Trail, 240, 244, 264, 265
Orofino, ID, 221
Outlaw Ridge, 262
Owen, Major John, 179

Pacific Fur Company, 235
Packer Meadows, 210
Palouse Falls, 231
Paradise Valley, MT, 149
Pasco, WA, 236
Patee House Museum, MO, 42

Paxson, Edgar Samuel, 203
Perry, Captain David, 217
Pi-Nee-Waus Center, 228
Pick-Sloan Project, 62, 108
Pierre's Hole, WY, 141
Pierre, SD, 66–70
Pike, Zebulon M., 47
Pilcher, Joshua, 141
Pillar Rock, 273
Pilot Peak, 147
Pine Ridge Indian Reservation, 75
Pioneer Courthouse Square, 265
Pioneer Mountains, 182
Pioneer Museum, MT, 110
Pipestone Creek, 159
Platte Purchase, 38
Platte River, NB, 47
Platte, SD, 63
Plummer, Henry, 162, 182
Point Ellice, 274
Point, Nicholas, 179
Polaris, MT, 182
Pomeroy, WA, 230
Pompeys Pillar, 144
Pony Express, 41
Poplar, MT, 104
Portland State University, 266
Portland, OR, 264–266
Pottawattamie County Jail, IA, 53
Potts, Pvt. John, 9, 89, 132, 135, 140
Powder River, 139
Powell Ranger Station, 211
Powell, John Wesley, 80
Prairie Creek, 169
Prairie Du Chien, Treaty of, 38
Prairie of the Knobs, 201
Prey, MT, 149
Price, General Sterling, 15, 30, 35
Price, Wilson Price, 222
Prickly Pear Creek, 128, 130
Prince of Wales Hotel, 193
Prior (Pryor), Sgt. Nathaniel, 9, 89, 140, 149
Provost, Etinenne, 141
Pryor Mountains, 144
Puget Sound, 277

Quantrill, Wm., 32, 35

Rainbow Falls, 127
Randle, WA, 262
Range Riders Museum, 139
Ravalli, Fr. Anthony, 179
Rawn, Capt. Charles, 206
Red Butte, 169
Red Lodge, MT, 146
Red Mountain, 201
Red River of the North, 92
Red Rock Lake, 170
Reed College, 266
Remington, Frederick, 126
Reno, Capt. Marcus, 142
Rhineland, MO, 21
Richland, WA, 236
Riggins, ID, 174
River of No Return (Salmon), 170, 173, 174
Riverdale, ND, 91
Robber's Roost, 162
Robidoux, Joseph, 40
Robinson Museum, SD, 69
Rocheport, MD, 28
Rock Creek Canyon, 147
Rocky Mountain Fur Co., 72, 101, 141
Rocky Mountains, 190
Rogers Pass, 200
Roosevelt Lodge, 148
Roosevelt, Franklin, 250
Roosevelt, Theodore, 204
Roosevelt, WA, 238
Rosburg, WA, 272
Rose, Edward (Cut Nose), 72
Ross's Hole, 176
Ross, Alexander, 177
Rowena Crest, OR, 248
Ruby River, 161, 163
Running Water, SD, 62
Russell Art Museum, 125
Russell, Charles M., 125, 131, 139, 177
Russell Majors and Waddell, 39
Ryan Dam Recreation Area, 123

Sacagawea (Sacajawea) (Sakakawea), 10, 73, 85, 86, 89, 90, 100, 109, 124, 127, 132, 134, 137, 164, 166, 172, 176, 183, 266, 283

Sacajewea State Park, 236
St. Albans, MO, 18
St. Charles, MO, 9, 16–17
St. Johns Bridge, 270
St. Joseph Indian School, SD, 65
St. Joseph Museum, MO, 42
St. Joseph, MO, 40–42
St. Louis, MO, 11–15
St. Louis Art Museum, 14
St. Louis Cathedral, 15
St. Marys Mission, 179
St. Louis Zoo, 14
Sakakawea State Park, ND, 91
Salmon, ID, 169, 173
Sandy River, 259, 263
Sanger, SD, 82
Santa Fe Trail, 17, 28, 32, 38
Sante Sioux Reservation, NB, 55
Sarpy County Museum, 48
Satus Pass, 238
Sauvie Island, 270
Schriver, Bob, 116, 193
Seaman, 10, 281
Seaside, OR, 282
Selby, SD, 71
Selway River, 216
7-Up Pete Creek, 201
Shannon, Pvt. John, 60, 89, 143
Shelby, MT, 190
Sheridan, Lt. Phil, 259
Sheridan, MT, 162
Sherman, General William T., 53, 91
Shields, Pvt. John, 22, 62, 88, 149, 169
Shoshoni Cove, 169
Shrine to Music Museum, SD, 60
Sidney, MT, 137, 138
Silver Gate, WY, 147, 148
Sinque Hole, 213
Sioux City, IA, 56, 57
Skamokawa, WA, 272
Slant Indian Village, SD, 78
Smith, Jedediah, 47, 72, 101, 135, 235
Smith, Joseph, 32, 33
Smoke Jumpers Center, 203
Snake River, 230
Snake/Columbia Confluence, 236
Snowbank Camp, 212

Soda Butte Creek, 147
Space, Ralph, 212
Spalding, Eliza, 227
Spalding, ID, 226
Spalding, Reverend Henry, 227
Spirit Mound, SD, 60
Spreckles, Alma de Bretville, 240
Springfield, SD, 62
Squaw Creek National Wildlife Refuge, MO, 44, 45
Standing Rock Indian Reservation, 73
Starlight Theatre, MO, 37
Stevens, Isaac, 101, 199, 220
Stevenson, WA, 259
Stevensville, MT, 179, 180
Stonehenge Monument, 240
Story, Nelson, 138
Sublette Bros, 141
Sublette, Bill, 15, 72
Sula, MT, 176
Sully, General Alfred H., 76
Sun River, 128, 198, 199
Surgeon's Quarters Museum, 244
Sweet Grass Hills, 187
Swope Park, MO, 37

Tanner Creek Falls, 257
Tavern Cave, MO, 19
Tendoy, ID, 171
Terry, Gen., 142
The Dalles Dam, 242
The Dallas, OR, 244, 248
Theodore Roosevelt National Park, 74, 96
Thespian Hall, 28
Thompson, David, 96
Thompson, Pvt. John, 89, 205
Thorn, Cpt. Jonathan, 276, 279
Three Forks, MT, 85, 131–137, 154, 155, 199
Thwaites, Dr. Reuben Gold, 1
Tillamook Head, 283
Timberline Lodge, 250
Tobacco Root Mountains, 136, 158
Toby, 173, 211, 212, 215
Tongue River, 139
Toohoolhoolzote, 227
Toppenish, 237

304 *Index*

Touchet River, 231
Trapper Peak, 177
Travelers Rest (Lolo), 128, 180, 186, 187, 204, 206
Treaty of Ghent, 192, 268
Troy, KS, 43
Truman Library, 32
Truman, Harry S, 32
Truteau, Baptiste, 12
Tsceminicum Park, 229
Tucamon Valley, 230
Twin Bridges, MT, 161, 176
Two Medicine Lakes, 195
Two Medicine River, 190

Umatilla, OR, 238
Union Pacific RR Museum, 50
Union Station Museum, NB, 50

Vancouver Barracks, 269
Vancouver, Cpt. Henry, 121
Vancouver, George, 241, 245
Vancouver, WA, 267–268
Vanderburgh, W. H., 141
Verendrye, Pierre Gaultier de Varennes, 69, 70, 82, 89
Vermillion, SD, 60
Vigilante Committee, 162, 182
Villard, Henry, 166
Virginia City, MT, 163

W. H. Over Museum, SD, 60
Wahkeena Falls, 258
Waiilatpu, WA, 233
Waitsburg, WA, 230, 232
Walker, Joseph, 255
Walking Coyote, 204
Walla Walla, WA, 232, 233
Wallula, WA, 199, 235
Walton, Izaak, 195
Warner, Pvt. Wm., 89, 205
Washburn, SD, 83, 86
Washington Park, Portland, 266
Washington, MO, 19, 20
Waterloo, MT, 159
Waterton Park, 193
Watford City, ND, 96
Waverly, MO, 29
Weippe Prairie, 215, 218–220

Weldon Spring Wildlife Area, MO, 18
West Glacier, 195
West Yellowstone, MT, 153
Western Heritage Center, MT, 146
Western Montana College, 166
Weston, MO, 37, 38
Westport Square, 36
Westport, Battle of, 35
Westport, MO, 34
White Cliffs, 117
White Cloud, KS, 43
White Salmon River, 251
White Salmon, WA, 251
White Swan, 238
White, John, 182
White Bird Battlefield, 216, 217
Whitehall, MT, 158
Whitehouse, Pvt. Joseph, 89
Whitehouse Pool, 212
Whitman College, 233
Whitman Mission, 233–235
Whitman, Dr. Marcus, 234, 235
Whitman, Narcissa, 235
Whoop-up Trail, 116, 197
Wilapa Bay, WA, 278
Wilkes, Captain, 277
Willamette Valley, 256, 259, 263
Williston, ND, 97, 98
Wilowa Valley, 227
Wind River, MT, 141
Wind River Indian Reservation, 86
Wind River, WA, 262
Windsor, Pvt. Richard, 89
Windy Ridge, 262
Winnebago Indian Reservation, 55
Winnipeg, Lake, 82
Wisdom, MT, 161, 183
Wiser, Pvt. Peter, 89
Wolf Point, MT, 105
Wood River, IL, 9
World Museum of Mining, 160
Wounded Knee, Battle of, 100
Wyeth, Nathaniel, 101, 235, 270

Yakima Nation Heritage Center, 237
Yakima Valley, 237
Yakima, WA, 238
Yankton, SD, 61, 62

Yeager, Erastus "Red," 162
Yellowstone Art Center, 145
Yellowstone County Museum, 145
Yellowstone Expedition, 53
Yellowstone National Park, 10, 141, 148
Yellowstone/Big Horn Confluence, 140
Yellowstone/Missouri Confluence, 98
York, 9, 71, 79
Young, Brigham, 33, 172
Young, John W., 166
Younger, John, 32